T0226076

FLORA OF THE
BRITISH ISLES

ILLUSTRATIONS

A. R. CLAPHAM   T. G. TUTIN   E. F. WARBURG

FLORA OF THE BRITISH ISLES

ILLUSTRATIONS

PART IV
MONOCOTYLEDONES

DRAWINGS BY
SYBIL J. ROLES

CAMBRIDGE
AT THE UNIVERSITY PRESS
1965

CAMBRIDGE UNIVERSITY PRESS
Cambridge, New York, Melbourne, Madrid, Cape Town, Singapore,
São Paulo, Delhi, Dubai, Tokyo, Mexico City

Cambridge University Press
The Edinburgh Building, Cambridge CB2 8RU, UK

Published in the United States of America by Cambridge University Press, New York

www.cambridge.org
Information on this title: www.cambridge.org/9780521269650

© Cambridge University Press 1965

This publication is in copyright. Subject to statutory exception
and to the provisions of relevant collective licensing agreements,
no reproduction of any part may take place without the written
permission of Cambridge University Press.

First published 1965
Re-issued 2010

A catalogue record for this publication is available from the British Library

ISBN 978-0-521-04661-9 Hardback
ISBN 978-0-521-26965-0 Paperback

Cambridge University Press has no responsibility for the persistence or
accuracy of URLs for external or third-party Internet Web sites referred to in
this publication, and does not guarantee that any content on such Web sites is,
or will remain, accurate or appropriate.

PREFACE

The fourth part of these Illustrations, consisting of the Monocotyledons, completes the set and follows the same general pattern as the earlier parts.

The sequence followed and the nomenclature adopted are those used in *Flora of the British Isles*, second edition, and the revised reprint of the *Excursion Flora*.

Miss Roles has completed the task of drawing nearly 2000 British plants, almost all from living specimens. We would like to take this opportunity of thanking her for the care she has displayed in making the drawings and for the patience and continued interest she has shown throughout the 15 years or so she has been doing them. We wish her a very happy retirement.

We are greatly indebted to the following for supplying many of the plants illustrated in this Part: Miss J. Allison, the late R. C. L. Burges, J. H. Chandler, Miss A. P. Conolly, G. I. Crawford, R. W. David, J. G. Dony, Miss U. K. Duncan, J. S. L. Gilmour, Mrs M. R. Gilson, the late R. A. Graham, G. Halliday, M. K. Hanson, C. B. Heginbotham, Miss J. E. Hibbard, E. K. Horwood, E. W. Jones, J. E. Lousley, D. McClintock, J. G. Packer, W. Palmer, Mrs M. S. Pennington, T. D. Pennington, C. D. Pigott, M. E. D. Poore, T. E. D. Poore, C. T. Prime, J. E. Raven, N. D. Simpson, Mrs F. Le Sueur, F. J. Taylor, J. Timson, S. M. Walters and Mrs E. W. Woodward.

A. R. C.
T. G. T.
E. F. W.

PART IV

MONOCOTYLEDONES

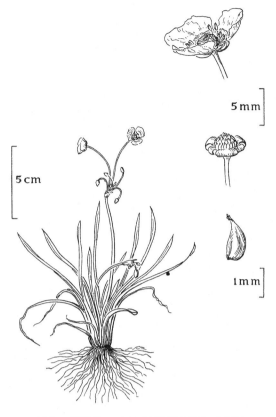

1454. *Baldellia ranunculoides* (L.) Parl.   Lesser
Water-Plantain   Pale purplish

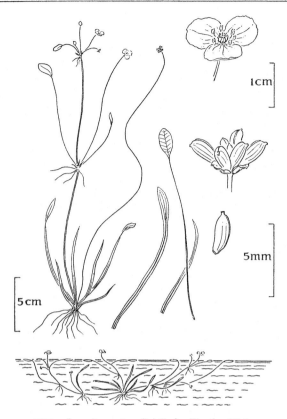

1455. *Luronium natans* (L.) Raf.   Floating Water-
Plantain   White and yellow

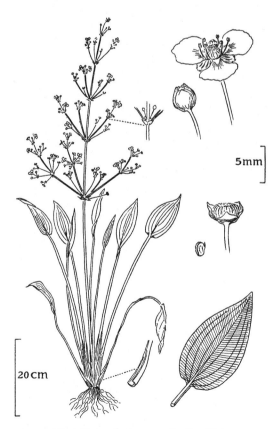

1456. *Alisma plantago-aquatica* L.   Water-
Plantain   Pale lilac

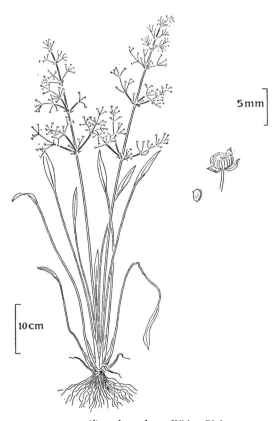

1457. *Alisma lanceolatum* With.   Pink

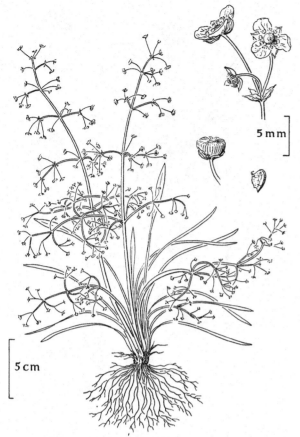

1458. *Alisma gramineum* Lejeune   Pale lilac

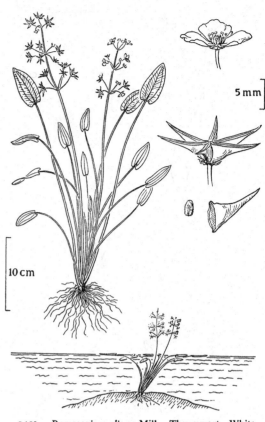

1459. *Damasonium alisma* Mill.   Thrumwort   White

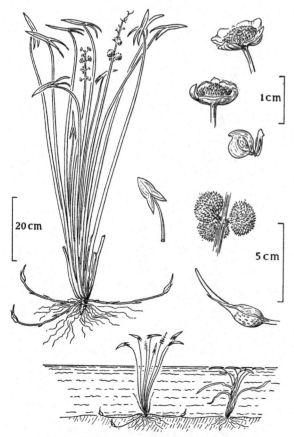

1460. *Sagittaria sagittifolia* L.   Arrow-head
White and dark lilac

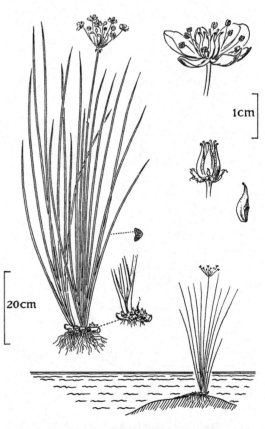

1461. *Butomus umbellatus* L.   Flowering Rush   Pink

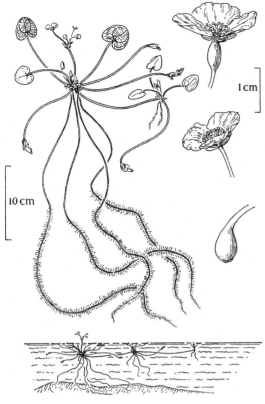

1462.   *Hydrocharis morsus-ranae* L.   Frog-bit
White and yellow

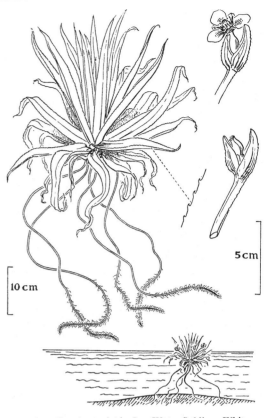

1463.   *Stratiotes aloides* L.   Water Soldier   White

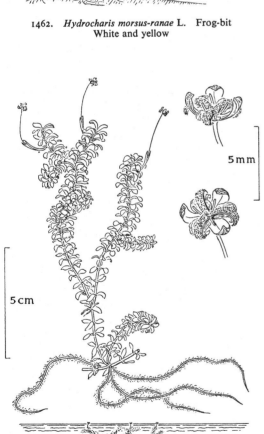

1464.   *Elodea canadensis* Michx.   Canadian Pondweed
Greenish-purple

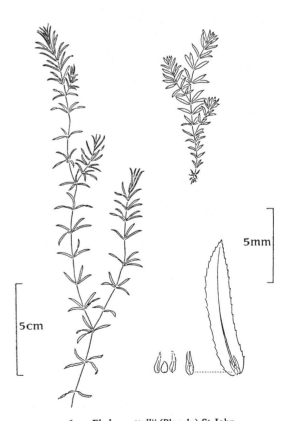

1465.   *Elodea nuttallii* (Planch.) St John

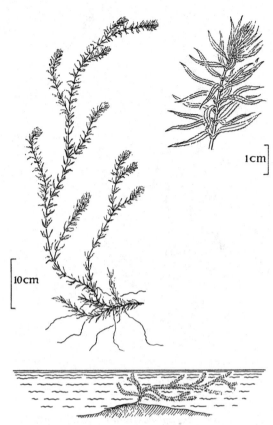

1466.   *Elodea callitrichoides* (Rich.) Casp.

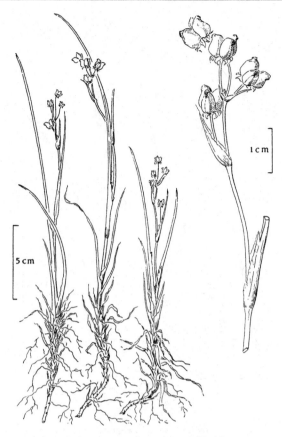

1467.   *Scheuchzeria palustris* L.   Yellowish-green

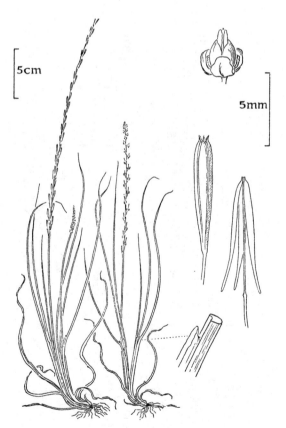

1468.   *Triglochin palustris* L.   Marsh Arrow-grass
         Green

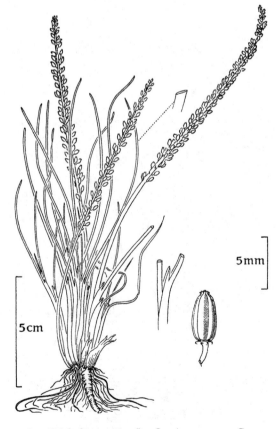

1469.   *Triglochin maritima* L.   Sea Arrow-grass   Green

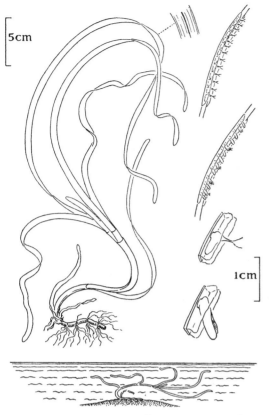

1470. *Zostera marina* L. Eel-grass, Grass-wrack
Green

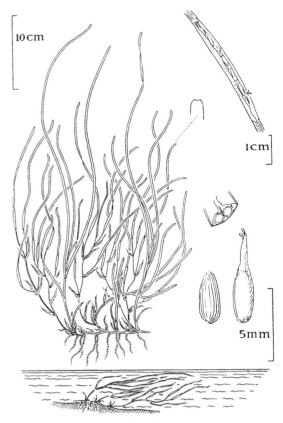

1471. *Zostera angustifolia* (Hornem.) Rchb. Green

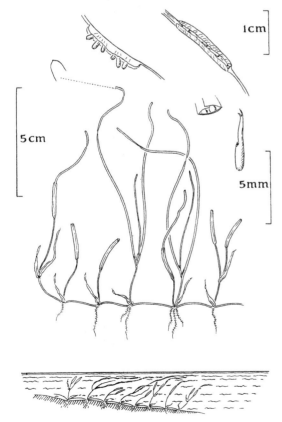

1472. *Zostera noltii* Hornem. Green

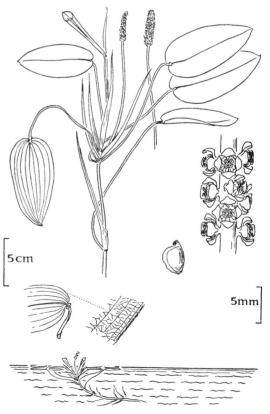

1473. *Potamogeton natans* L. Broad-leaved Pondweed

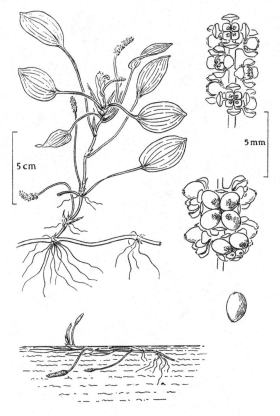

1474.   *Potamogeton polygonifolius* Pourr.   Bog Pondweed

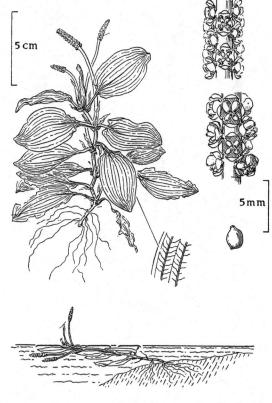

1475.   *Potamogeton coloratus* Hornem.   Fen Pondweed

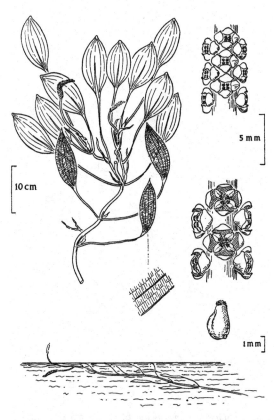

1476.   *Potamogeton nodosus* Poir.   Loddon Pondweed

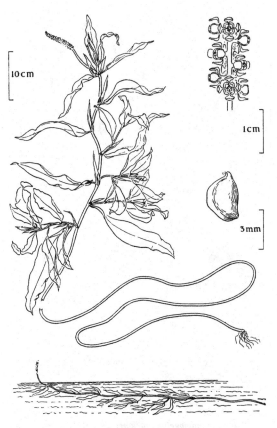

1477.   *Potamogeton lucens* L.   Shining Pondweed

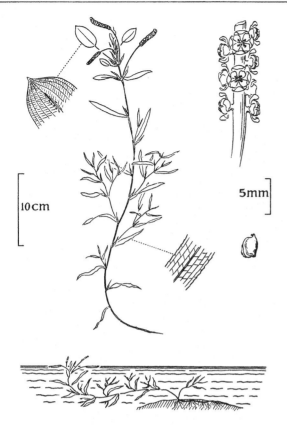

1478. *Potamogeton gramineus* L.   Various-leaved Pondweed

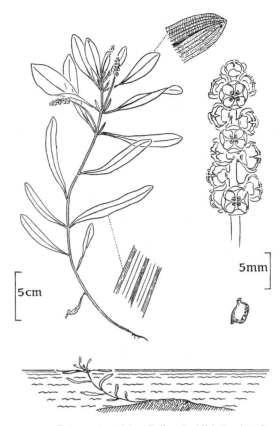

1479. *Potamogeton alpinus* Balb.   Reddish Pondweed

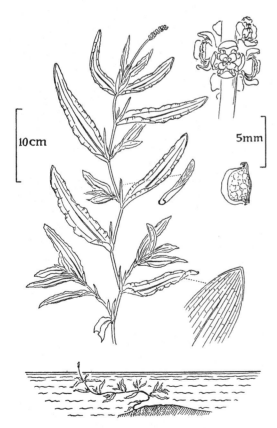

1480. *Potamogeton praelongus* Wulf.   Long-
stalked Pondweed

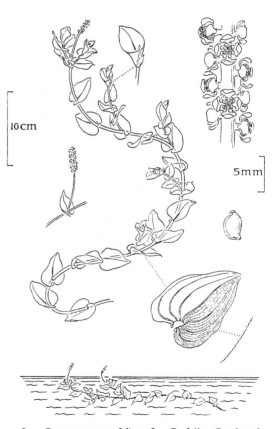

1481. *Potamogeton perfoliatus* L.   Perfoliate Pondweed

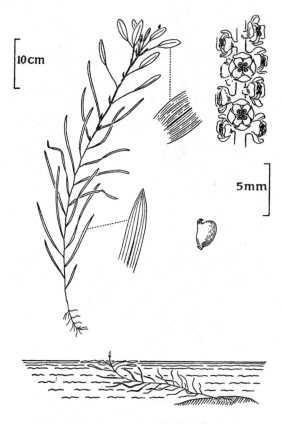

1482. *Potamogeton epihydrus* Raf.   Leafy Pondweed

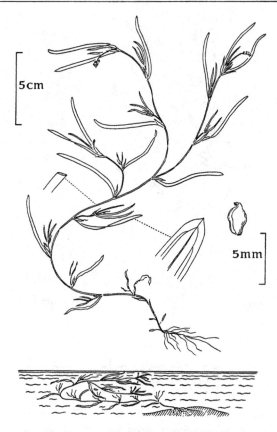

1483. *Potamogeton friesii* Rupr.   Flat-stalked Pondweed

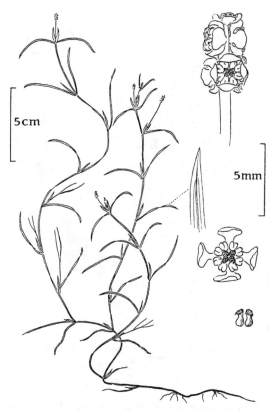

1484. *Potamogeton rutilus* Wolfg.   Shetland Pondweed

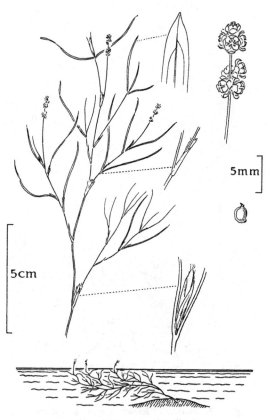

1485. *Potamogeton pusillus* L.

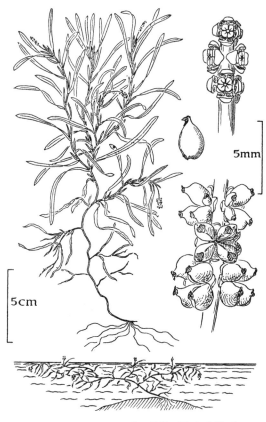

1486.   *Potamogeton obtusifolius* Mert. & Koch
Grassy Pondweed

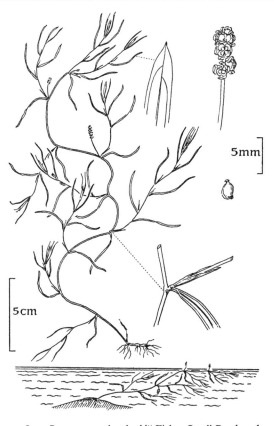

1487.   *Potamogeton berchtoldii* Fieb.   Small Pondweed

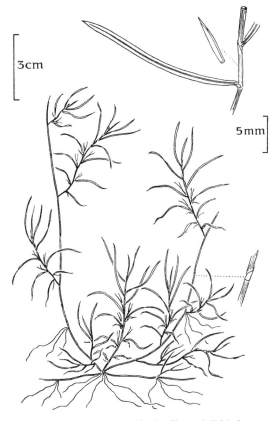

1488.   *Potamogeton trichoides* Cham. & Schlecht.
Hair-like Pondweed

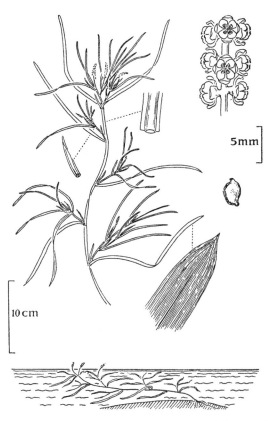

1489.   *Potamogeton compressus* L.   Grass-wrack Pondweed

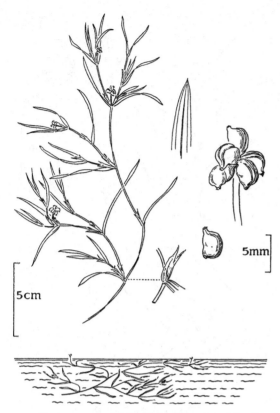

1490. *Potamogeton acutifolius* Link   Sharp-leaved Pondweed

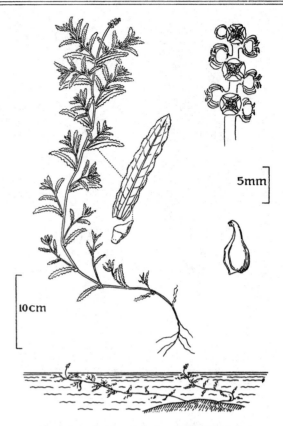

1491. *Potamogeton crispus* L.   Curled Pondweed

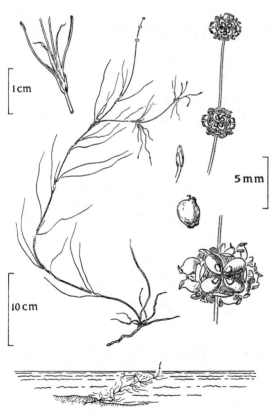

1492. *Potamogeton filiformis* Pers.   Slender-leaved Pondweed

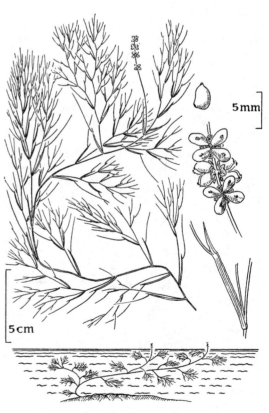

1493. *Potamogeton pectinatus* L.   Fennel-leaved Pondweed

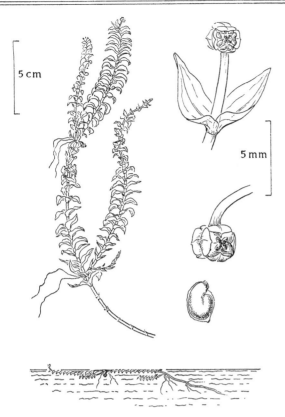

1494.   *Groenlandia densa* (L.) Fourr.   Opposite-
leaved Pondweed

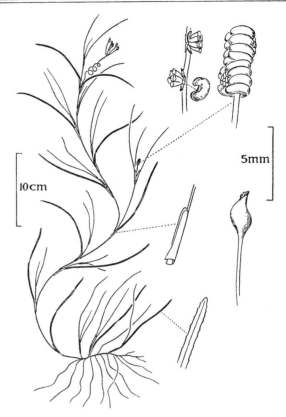

1495.   *Ruppia spiralis* L. ex Dumort.

1496.   *Ruppia maritima* L.

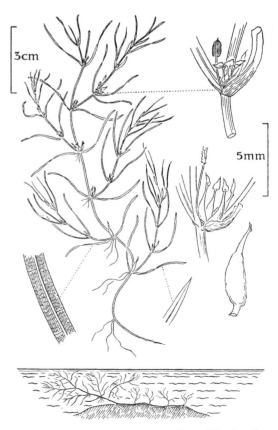

1497.   *Zannichellia palustris* L.   Horned Pondweed

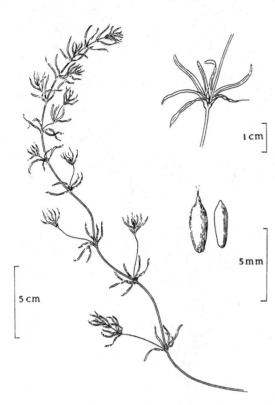

1498. *Najas flexilis* (Willd.) Rostk. & Schmidt

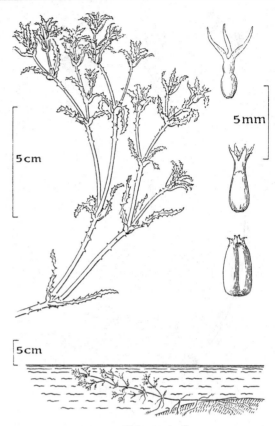

1499. *Najas marina* L.

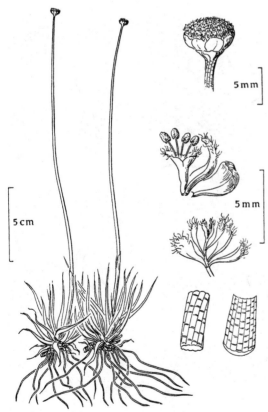

1500. *Eriocaulon septangulare* With.   Pipe-wort
Greyish

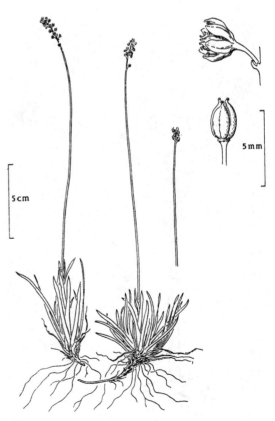

1501. *Tofieldia pusilla* (Michx.) Pers.   Scottish Asphodel
Greenish-white

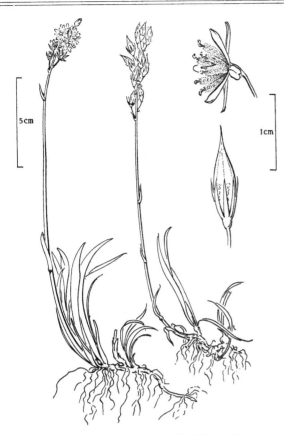

1502. *Narthecium ossifragum* (L.) Huds. Bog
Asphodel Yellow

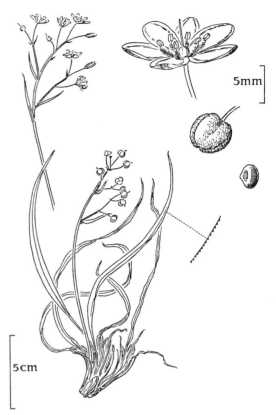

1503. *Simethis planifolia* (L.) Gren. & Godr. White
and purplish

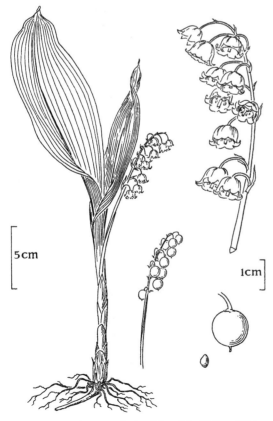

1504. *Convallaria majalis* L. Lily-of-the-Valley White

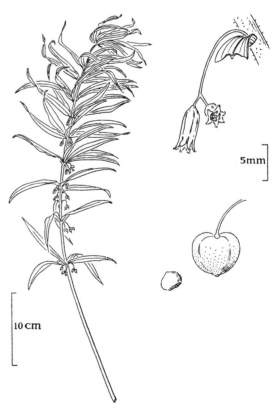

1505. *Polygonatum verticillatum* (L.) All. Whorled
Solomon's Seal Greenish-white

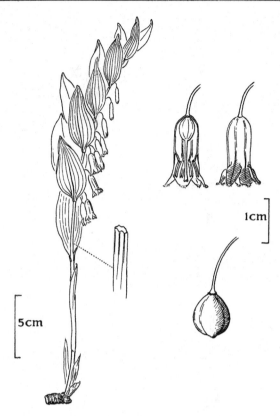

1506. *Polygonatum odoratum* (Mill.) Druce   Angular
Solomon's Seal   Greenish-white

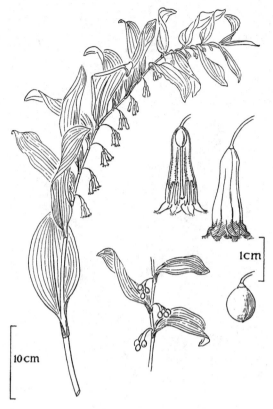

1507. *Polygonatum multiflorum* (L.) All.   Solomon's Seal
Greenish-white

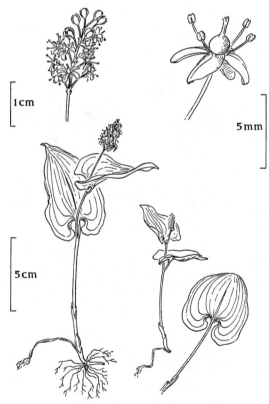

1508. *Maianthemum bifolium* (L.) Schmidt   May Lily
White

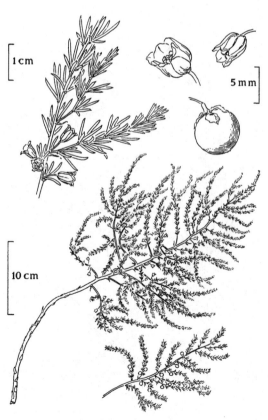

1509. *Asparagus officinalis* ssp. *prostratus* (Dumort.)
E. F. Warb.   Asparagus   Greenish

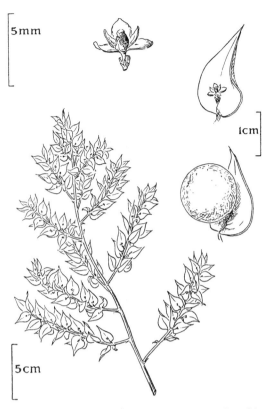

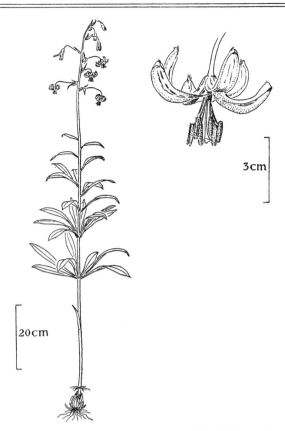

1510. *Ruscus aculeatus* L. Butcher's Broom Greenish

1511. *Lilium martagon* L. Martagon Lily Dull purple

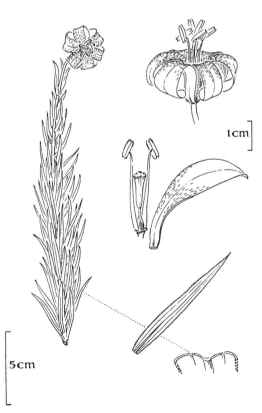

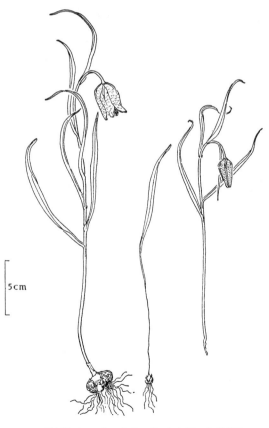

1512. *Lilium pyrenaicum* Gouan Yellow and black

1513. *Fritillaria meleagris* L. Snake's Head, Fritillary
Chequered purple

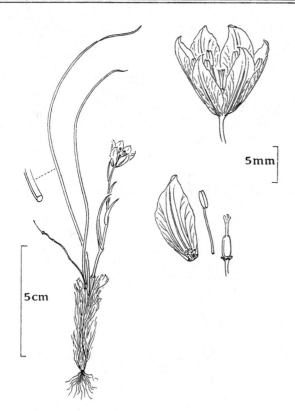

1514. *Lloydia serotina* (L.) Rchb.   Lloydia   White,
purple-veined

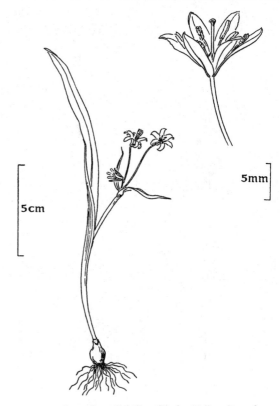

1515. *Gagea lutea* (L.) Ker.-Gawl.   Yellow Star-of-
Bethlehem   Yellow

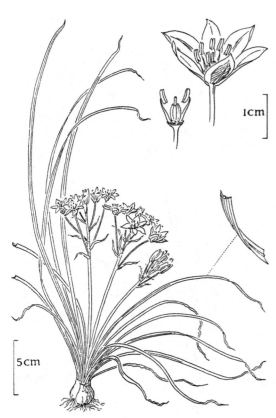

1516. *Ornithogalum umbellatum* L.   Star-of-Bethlehem
White

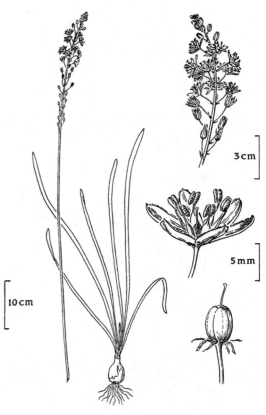

1517. *Ornithogalum pyrenaicum* L.   Bath Asparagus
Greenish-white

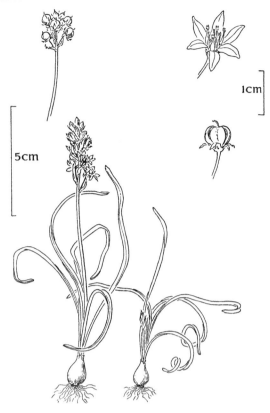

1518.   *Scilla verna* Huds.   Spring Squill   Violet-blue

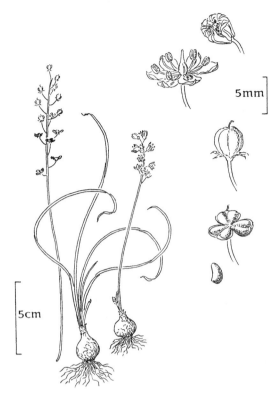

1519   *Scilla autumnalis* L.   Autumnal Squill   Purple

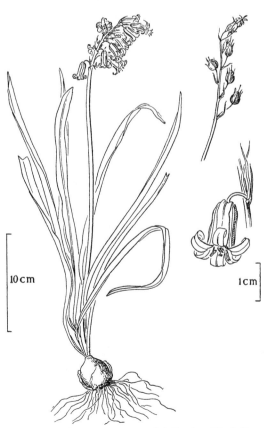

1520.   *Endymion non-scriptus* (L.) Garcke   Bluebell,
Wild Hyacinth   Violet-blue

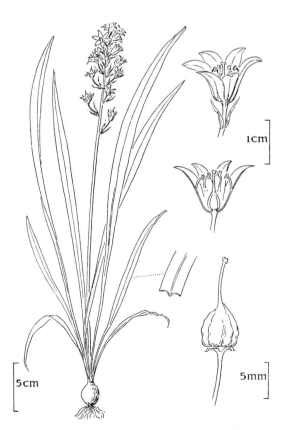

1521.   *Endymion hispanicus* (Mill.) Chouard   Blue

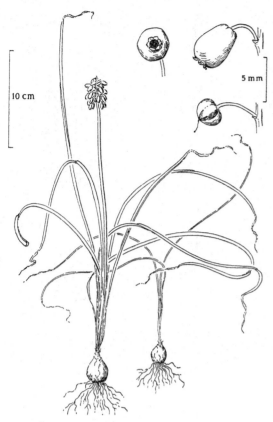

1522. *Muscari atlanticum* Boiss. & Reut. Grape Hyacinth
Dark blue

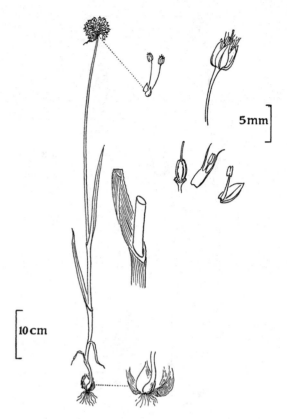

1523. *Allium ampeloprasum* L. Wild Leek Pale purple

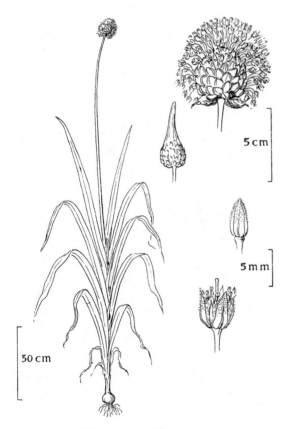

1524. *Allium babingtonii* Borrer Pale purple

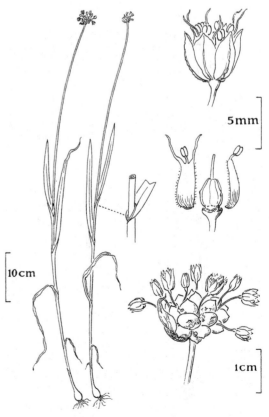

1525. *Allium scorodoprasum* L. Sand Leek Reddish-
purple

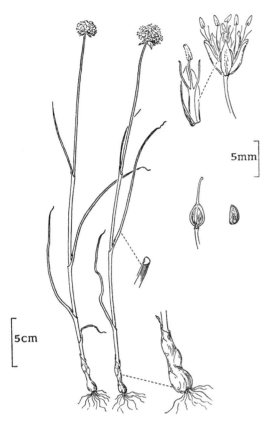

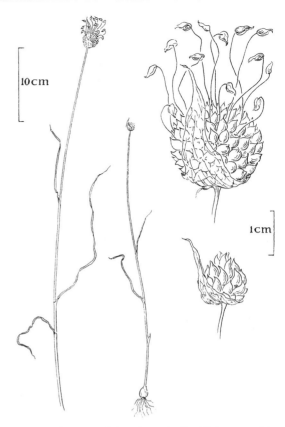

1526.   *Allium sphaerocephalon* L.   Round-headed Leek
Reddish-purple

1527.   *Allium vineale* L.   Crow Garlic   Pink or greenish-
white

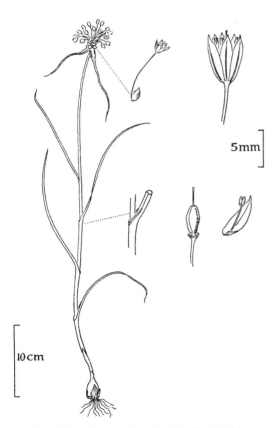

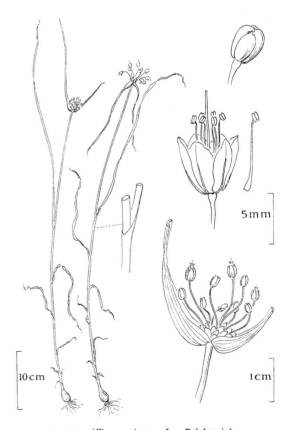

1528.   *Allium oleraceum* L.   Field Garlic   Pinkish or
greenish

1529.   *Allium carinatum* L.   Bright pink

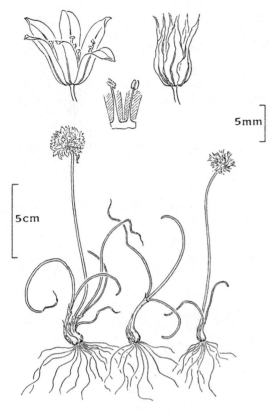

1530. *Allium schoenoprasum* L.   Chives   Purple or pink

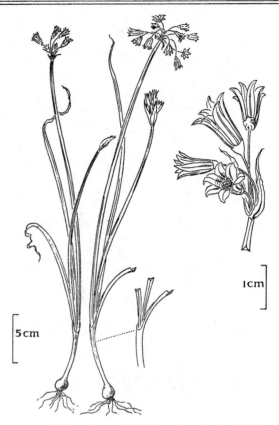

1531. *Allium triquetrum* L.   Triquetrous Garlic   White

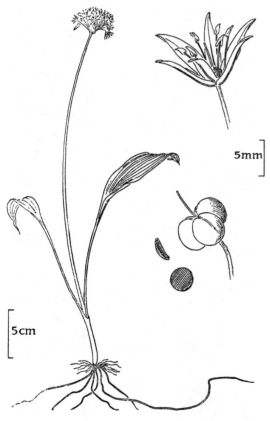

1532. *Allium ursinum* L.   Ramsons   White

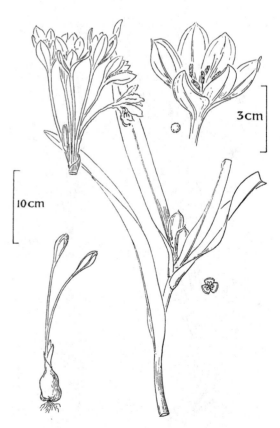

1533. *Colchicum autumnale* L.   Meadow Saffron, Autumn
Crocus   Pale purple

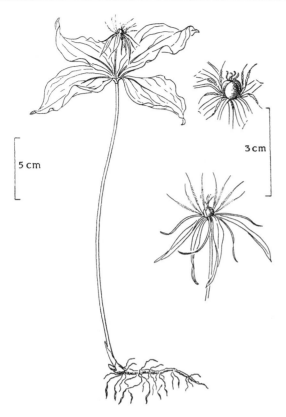

1534. *Paris quadrifolia* L.   Herb Paris   Green

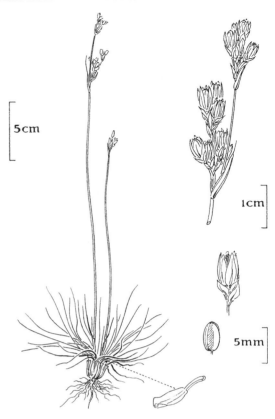

1535. *Juncus squarrosus* L.   Heath Rush   Dark brown

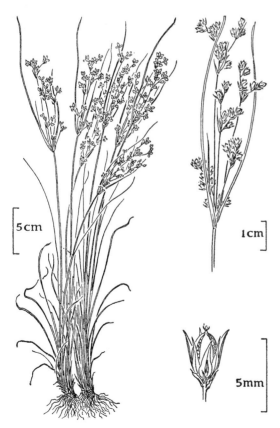

1536. *Juncus tenuis* Willd.   Greenish

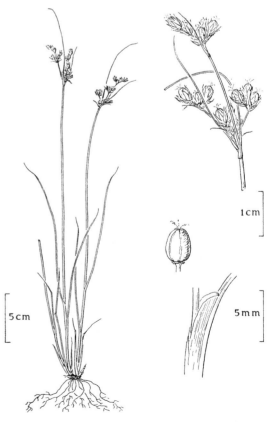

1537. *Juncus dudleyi* Wiegand   Greenish

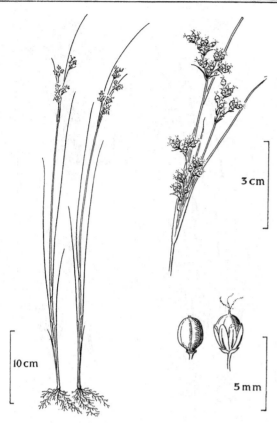

1538.   *Juncus compressus* Jacq.   Round-fruited Rush
Light brown

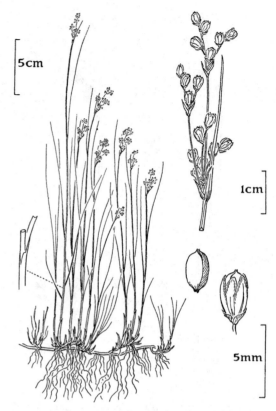

1539.   *Juncus gerardi* Lois.   Mud Rush   Dark brown

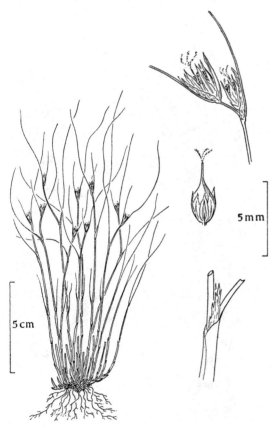

1540.   *Juncus trifidus* L.   Three-leaved Rush   Dark brown

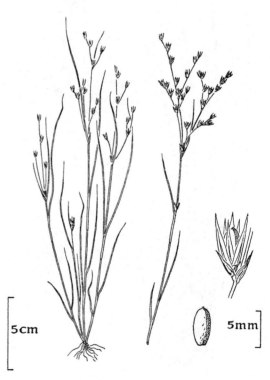

1541.   *Juncus bufonius* L.   Toad Rush   Pale green

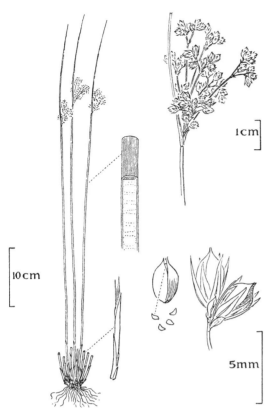

1542.   *Juncus inflexus* L.   Hard Rush   Brown

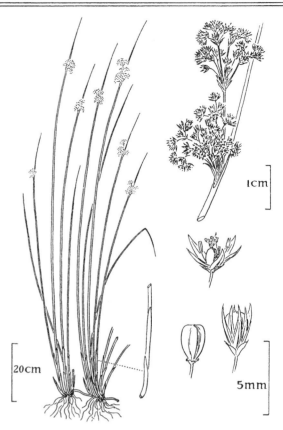

1543.   *Juncus effusus* L.   Soft Rush   Brown

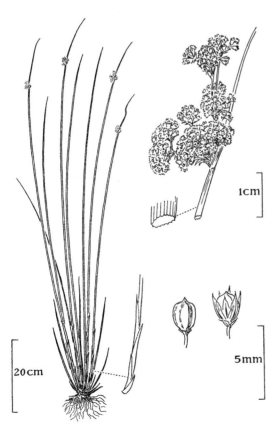

1544.   *Juncus conglomeratus* L.   Brown

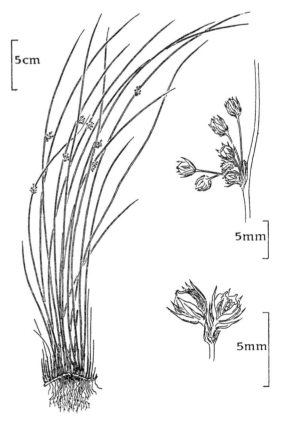

1545.   *Juncus filiformis* L.   Straw-coloured

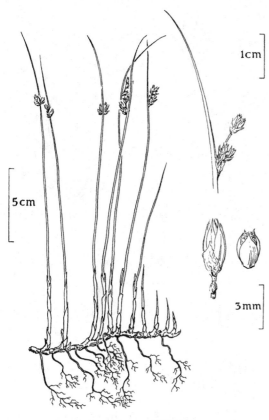

1546. *Juncus balticus* Willd.   Brown

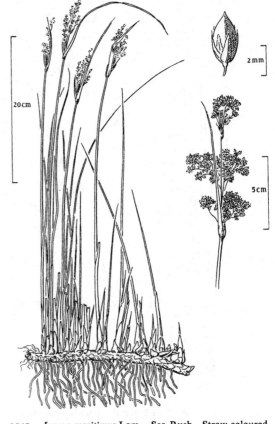

1547. *Juncus maritimus* Lam.   Sea Rush   Straw-coloured

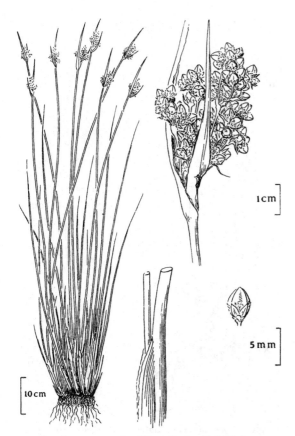

1548. *Juncus acutus* L.   Sharp Rush   Reddish-brown

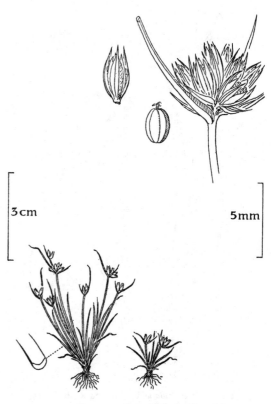

1549. *Juncus capitatus* Weigel   Greenish

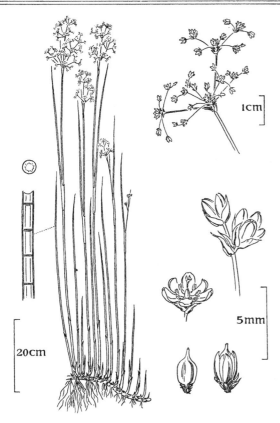

1550. *Juncus subnodulosus* Schrank   Blunt-flowered Rush
         Brown

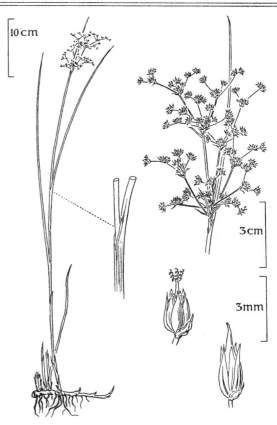

1551. *Juncus acutiflorus* Ehrh. ex Hoffm.   Sharp-flowered
         Rush   Brown

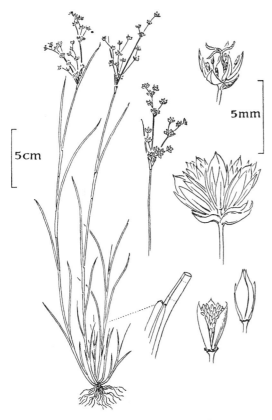

1552. *Juncus articulatus* L.   Jointed Rush   Brown

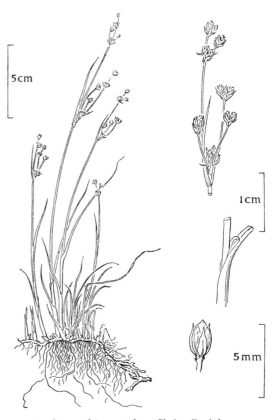

1553. *Juncus alpinoarticulatus* Chaix   Dark brown or
         blackish

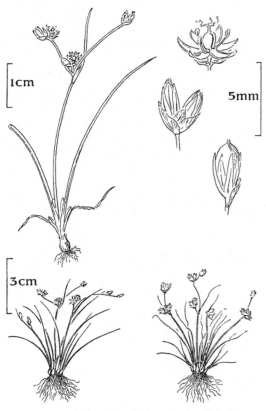

1554. *Juncus bulbosus* L.   Bulbous Rush   Light brown

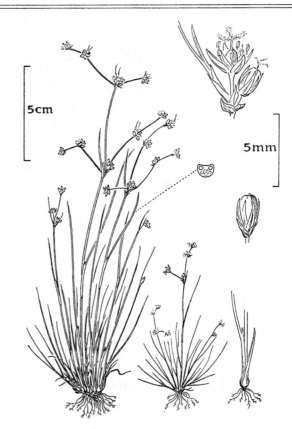

1555. *Juncus kochii* F. W. Schultz   Dark brown

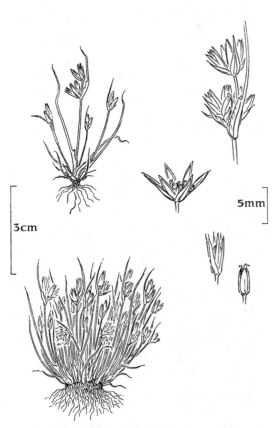

1556. *Juncus mutabilis* Lam.   Dwarf Rush   Greenish

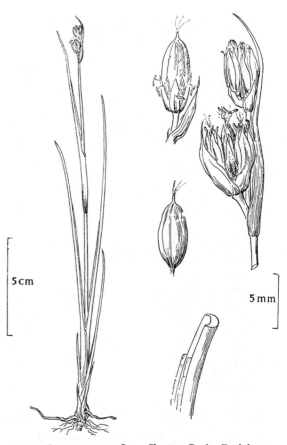

1557. *Juncus castaneus* Sm.   Chestnut Rush   Dark brown

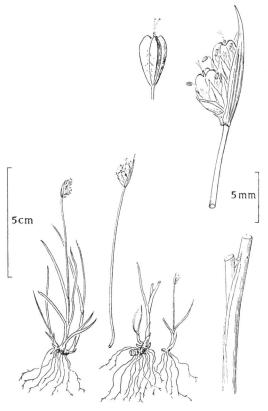

1558. *Juncus biglumis* L.   Two-flowered Rush   Purplish-brown

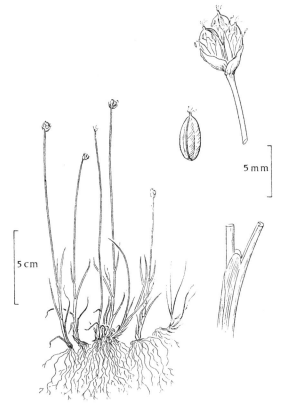

1559. *Juncus triglumis* L.   Three-flowered Rush   Light brown

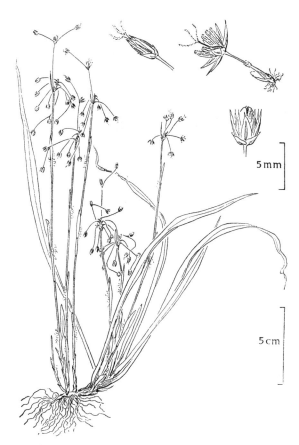

1560. *Luzula pilosa* (L.) Willd.   Hairy Woodrush   Dark brown

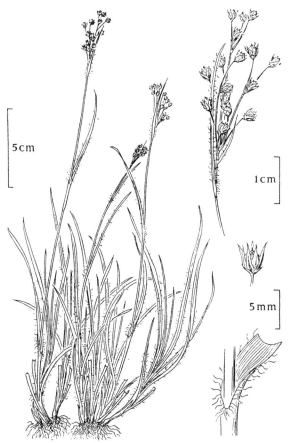

1561. *Luzula forsteri* (Sm.) DC.   Forster's Woodrush   Reddish-brown

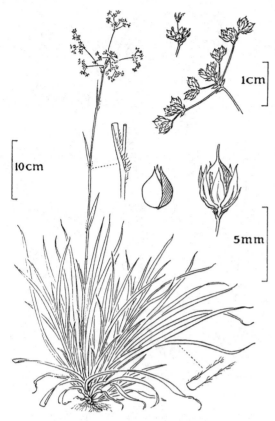

1562.   *Luzula sylvatica* (Huds.) Gaud.   Greater Woodrush
Chestnut-brown

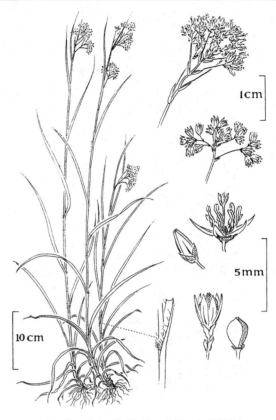

1563.   *Luzula luzuloides* (Lam.) Dandy & Wilmott
Dirty white

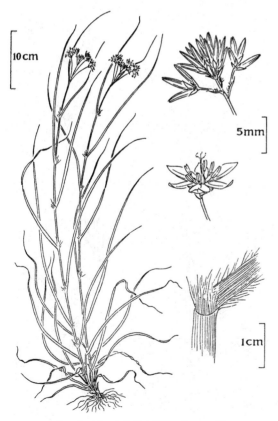

1564.   *Luzula nivea* (L.) DC.   Snow-white

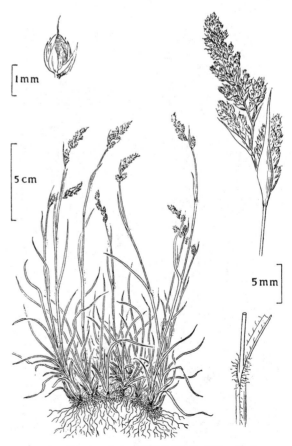

1565.   *Luzula spicata* (L.) DC.   Spiked Woodrush
Chestnut-brown

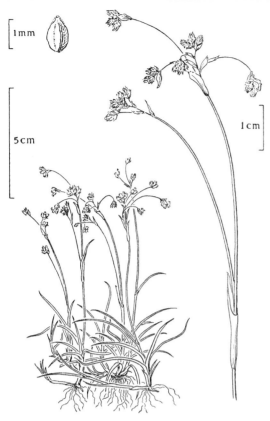

1566. *Luzula arcuata* Sw. Curved Woodrush Brown

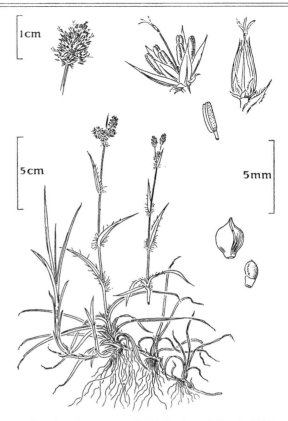

1567. *Luzula campestris* (L.) DC. Sweep's Brush, Field
Woodrush Chestnut-brown

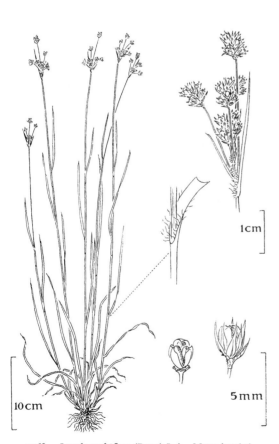

1568. *Luzula multiflora* (Retz.) Lej. Many-headed
Woodrush Chestnut-brown

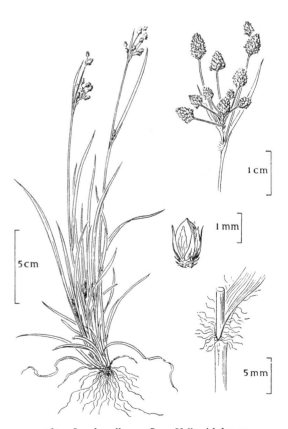

1569. *Luzula pallescens* Sw. Yellowish-brown

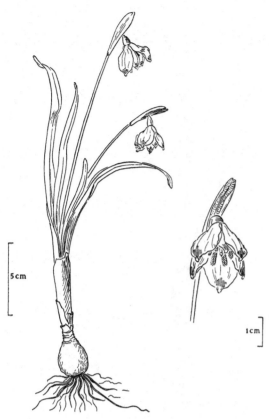

1570. *Leucojum vernum* L. Spring Snowflake White
and green

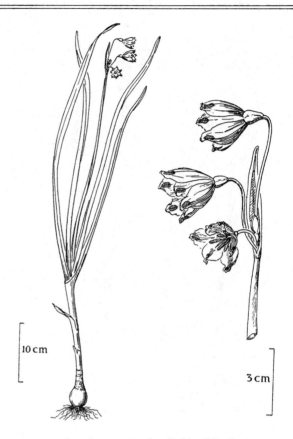

1571. *Leucojum aestivum* L. Loddon Lily, Summer
Snowflake White and green

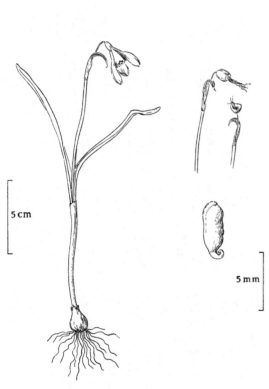

1572. *Galanthus nivalis* L. Snowdrop White and green

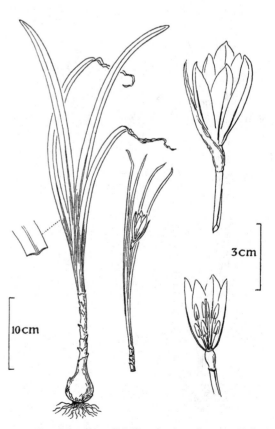

1573. *Sternbergia lutea* (L.) Ker.-Gawl. ex Spreng. Yellow

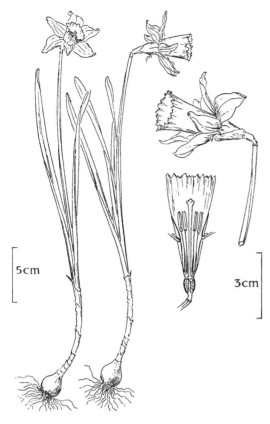

1574. *Narcissus pseudonarcissus* L. Wild Daffodil Yellow

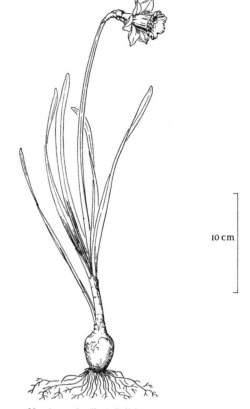

1575. *Narcissus obvallaris* Salisb. Tenby Daffodil Yellow

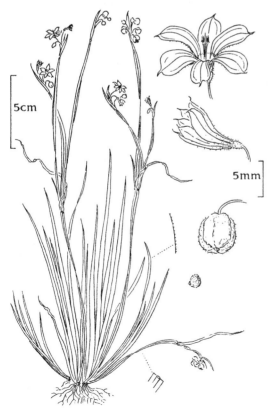

1576. *Sisyrinchium bermudiana* L. Blue-eyed Grass Blue

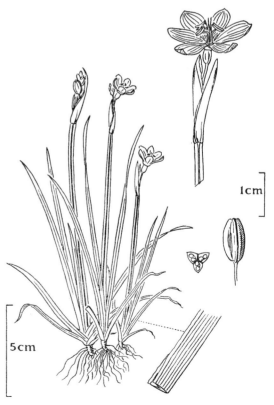

1577. *Sisyrinchium californicum* (Ker.-Gawl.) Ait.f. Yellow

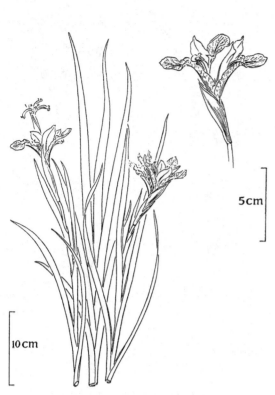

1578. *Iris spuria* L.   Blue-violet

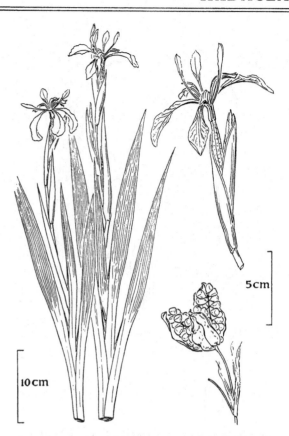

1579. *Iris foetidissima* L.   Gladdon, Stinking Iris
Purplish-livid

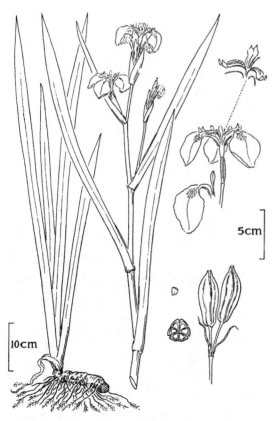

1580. *Iris pseudacorus* L.   Yellow Flag   Yellow

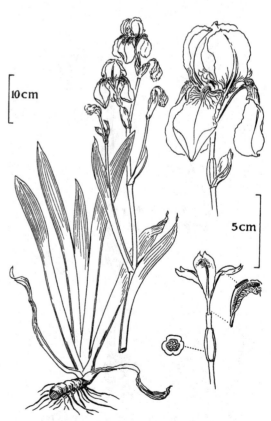

1581. *Iris germanica* L.   Dark purple

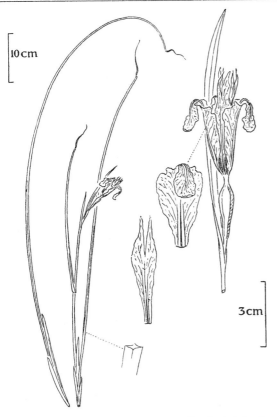

1582. *Hermodactylus tuberosus* (L.) Mill. Snake's-head
Iris   Smoky purple and greenish-yellow

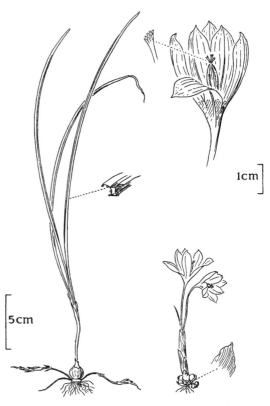

1583. *Crocus nudiflorus* Sm.   Autumnal Crocus
Pale purple

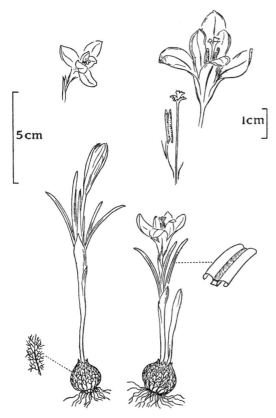

1584. *Crocus purpureus* Weston   Purple Crocus
Purple or white

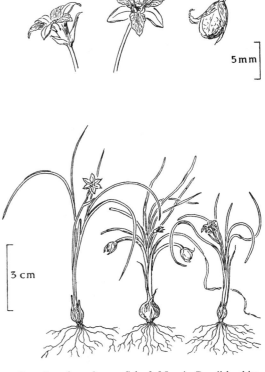

1585. *Romulea columnae* Seb. & Mauri   Purplish-white

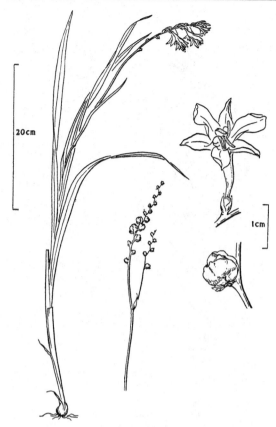

1586.   *Crocosmia × crocosmiflora* (Lemoine) N.E.Br.
Montbretia   Deep orange

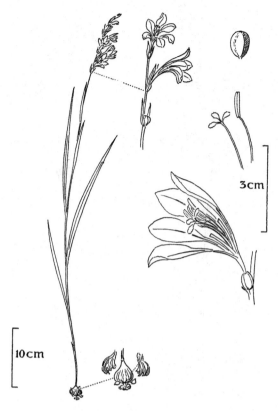

1587.   *Gladiolus illyricus* Koch   Crimson-purple

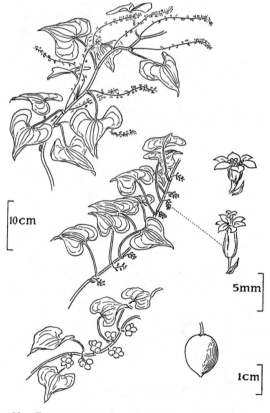

1588.   *Tamus communis* L.   Black Bryony   Yellowish-green

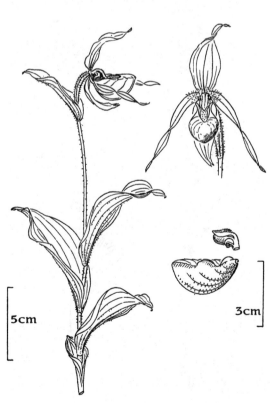

1589.   *Cypripedium calceolus* L.   Lady's Slipper   Maroon
and pale yellow

[35]

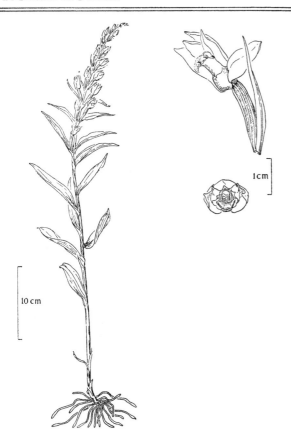

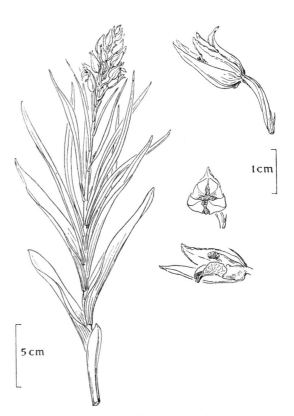

1590.   *Cephalanthera damasonium* (Mill.) Druce   White
Helleborine   White

1591.   *Cephalanthera longifolia* (L.) Fritsch   Long-leaved
Helleborine   White

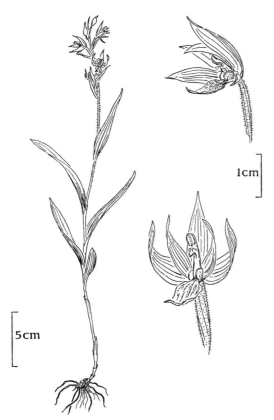

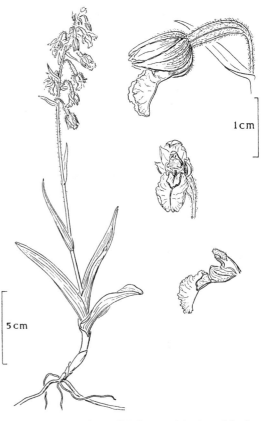

1592.   *Cephalanthera rubra* (L.) Rich.   Red Helleborine
Red

1593.   *Epipactis palustris* (L.) Crantz   Marsh Helleborine
Purplish-green and white

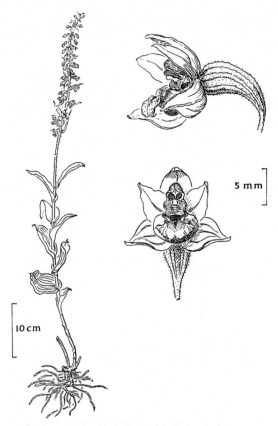

1594. *Epipactis helleborine* (L.) Crantz   Broad Helleborine
Greenish to dull purple

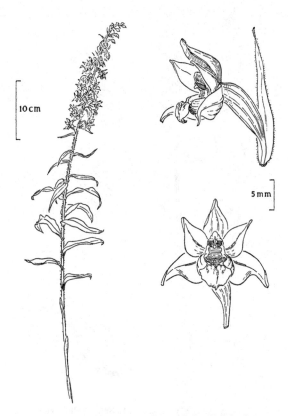

1595. *Epipactis purpurata* Sm.   Violet Helleborine
Greenish-white

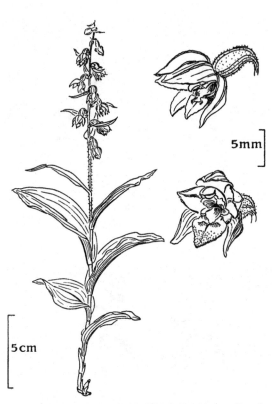

1596. *Epipactis leptochila* (Godf.) Godf.   Narrow-lipped
Helleborine   Yellow-green

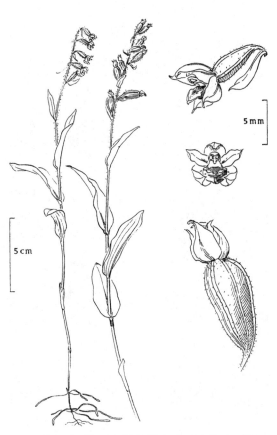

1597. *Epipactis dunensis* (T. & T. A. Steph.) Godf.   Dune
Helleborine   Yellow-green

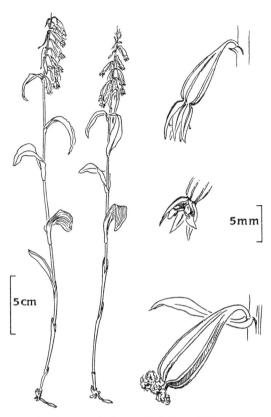

1598. *Epipactis phyllanthes* G. E. Sm. Yellow-green

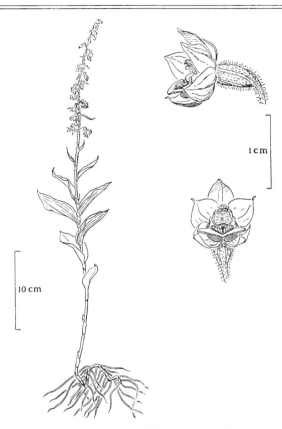

1599. *Epipactis atrorubens* (Hoffm.) Schult. Dark-red
Helleborine Red-purple

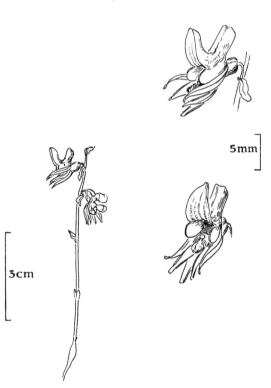

1600. *Epipogium aphyllum* Sw. Ghost Orchid Yellow
or reddish

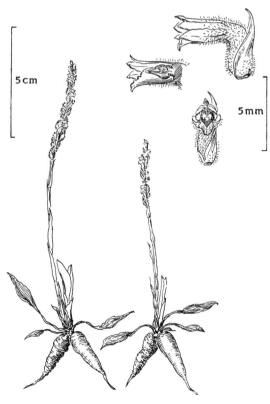

1601. *Spiranthes spiralis* (L.) Chevall. Autumn Lady's
Tresses White

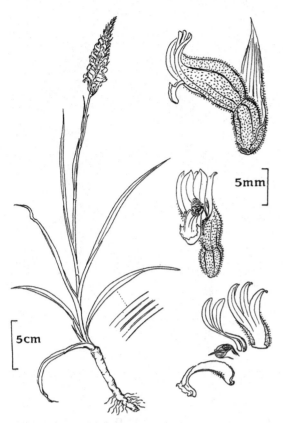

1602.   *Spiranthes romanzoffiana* Cham.   Drooping Lady's
Tresses   White

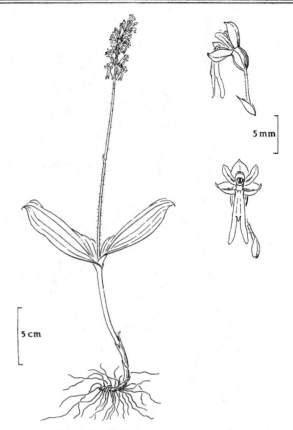

1603.   *Listera ovata* (L.) R.Br.   Twayblade   Yellow-green

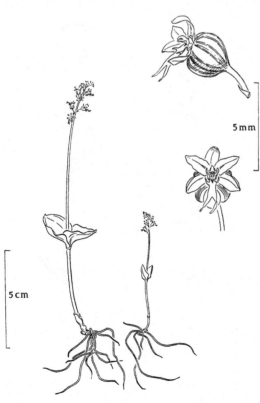

1604.   *Listera cordata* (L.) R.Br.   Lesser Twayblade
Reddish-green

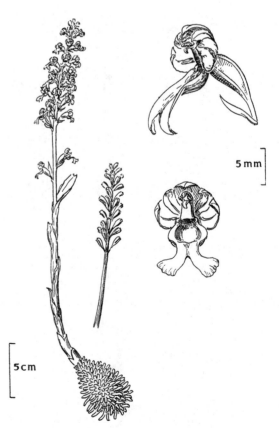

1605.   *Neottia nidus-avis* (L.) Rich.   Bird's-nest Orchid
Brown

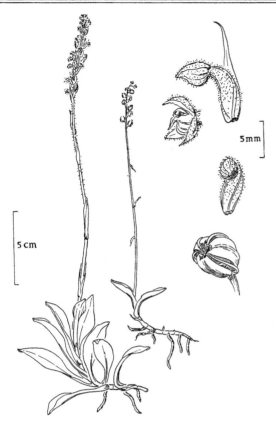

1606.   *Goodyera repens* (L.) R.Br.   Creeping Lady's Tresses
Cream

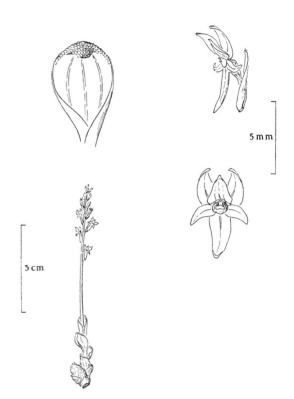

1607.   *Hammarbya paludosa* (L.) O. Kuntze   Bog Orchid
Yellow-green

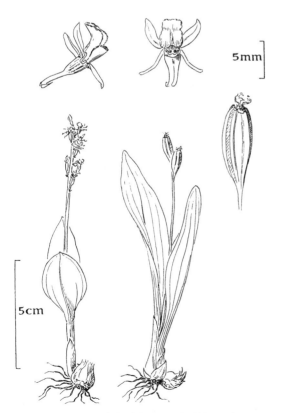

1608.   *Liparis loeselii* (L.) Rich.   Fen Orchid   Yellow-green

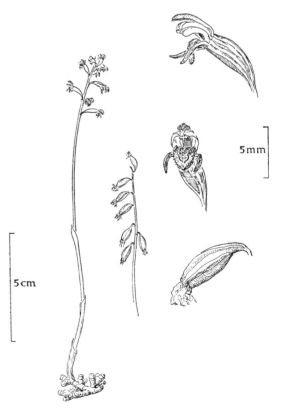

1609.   *Corallorhiza trifida* Chatel.   Coral-root   Greenish

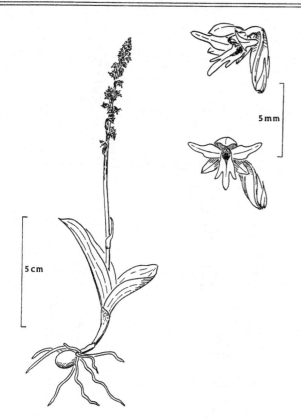

1610. *Herminium monorchis* (L.) R.Br. Musk Orchid
Greenish

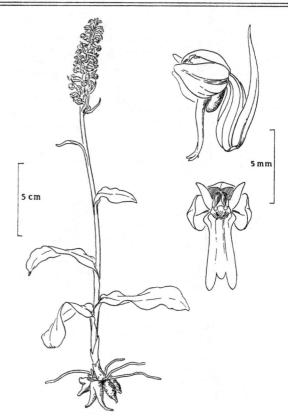

1611. *Coeloglossum viride* (L.) Hartm. Frog Orchid
Greenish

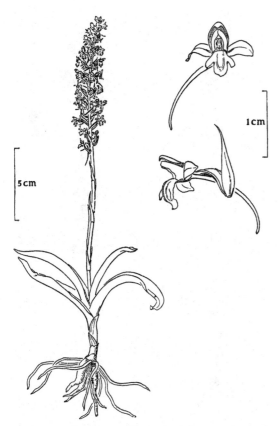

1612. *Gymnadenia conopsea* (L.) R.Br. Fragrant Orchid
Reddish-lilac

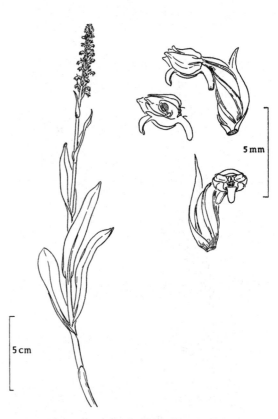

1613. *Leuchorchis albida* (L.) E. Mey. Small White
Orchid Greenish-white

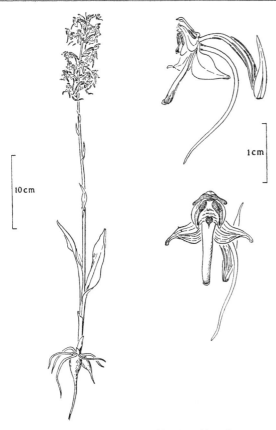

1614.   *Platanthera chlorantha* (Cust.) Rchb.   Greater
Butterfly Orchid   Greenish-white

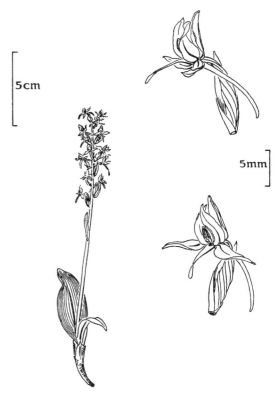

1615.   *Platanthera bifolia* (L.) Rich.   Lesser Butterfly
Orchid   Whitish

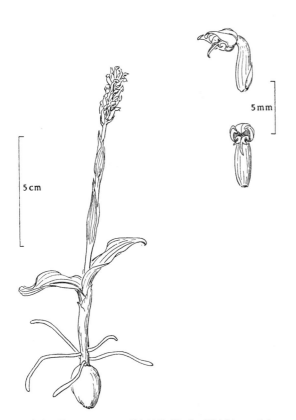

1616.   *Neotinea intacta* (Link) Rchb. f.   Whitish or pink

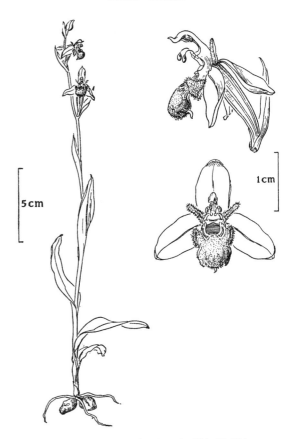

1617.   *Ophrys apifera* Huds.   Bee Orchid   Pinkish-green
and purple

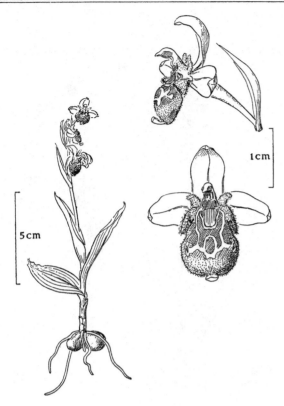

1618.   *Ophrys fuciflora* (Crantz) Moench   Late Spider
Orchid   Pinkish and maroon

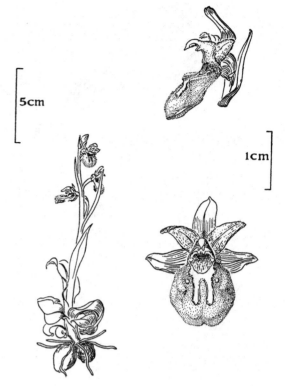

1619.   *Ophrys sphegodes* Mill.   Early Spider Orchid
Yellow-green and purple-brown

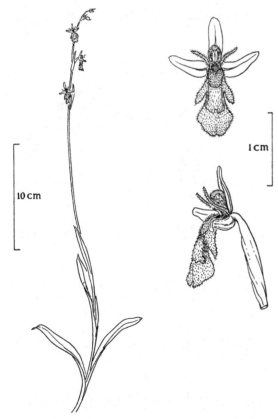

1620.   *Ophrys insectifera* L.   Fly Orchid   Yellow-green,
purple-brown and bluish

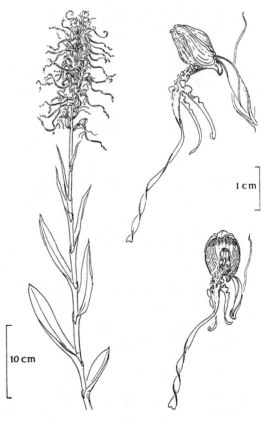

1621.   *Himantoglossum hircinum* (L.) Spreng.   Lizard
Orchid   Greenish

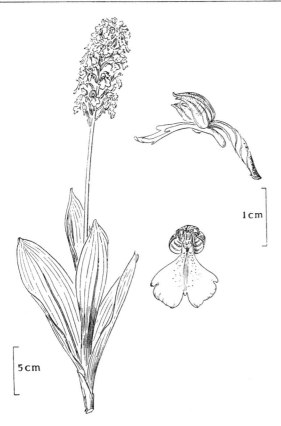

1622. *Orchis purpurea* Huds.   Lady Orchid   Reddish-
purple

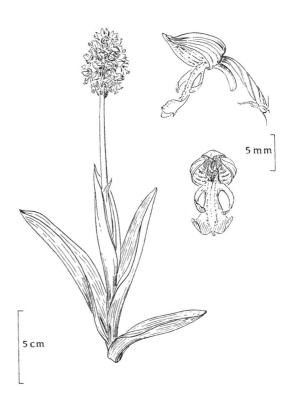

1623. *Orchis militaris* L.   Soldier Orchid   Reddish-violet

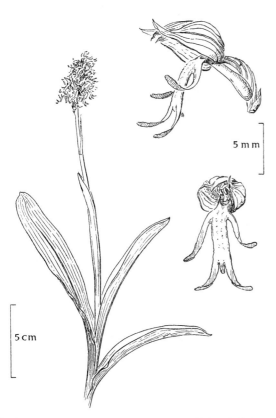

1624. *Orchis simia* Lam.   Monkey Orchid   Whitish,
pink and violet

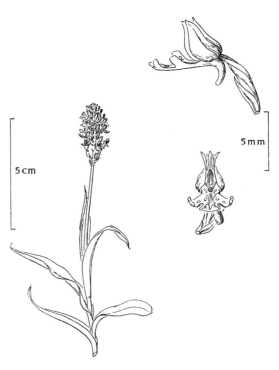

1625. *Orchis ustulata* L.   Dark-winged Orchid, Burnt
Orchid   Dark maroon, becoming whitish

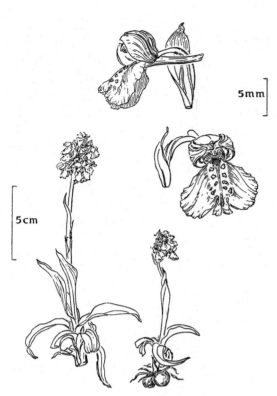

1626. *Orchis morio* L.   Green-winged Orchid   Reddish-purple

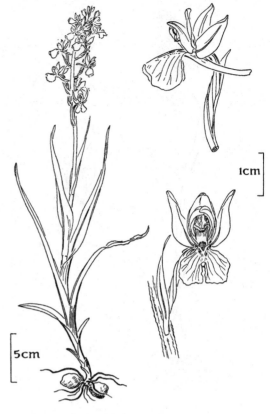

1627 *Orchis laxiflora* Lam.   Jersey Orchid   Dark reddish-purple

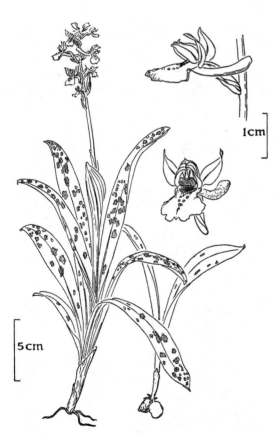

1628. *Orchis mascula* (L.) L.   Early Purple Orchid, Blue Butcher   Purplish-crimson

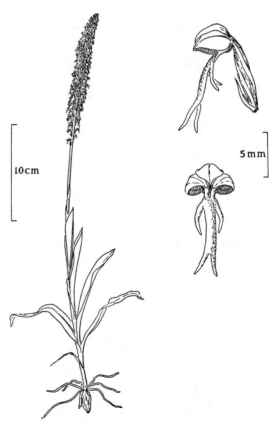

1629. *Aceras anthropophorum* (L.) Ait. f.   Man Orchid Greenish-yellow

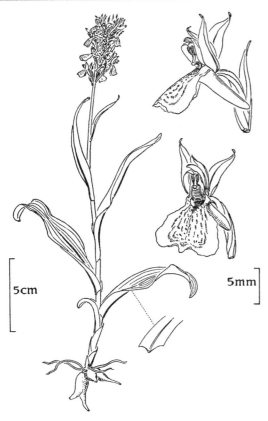

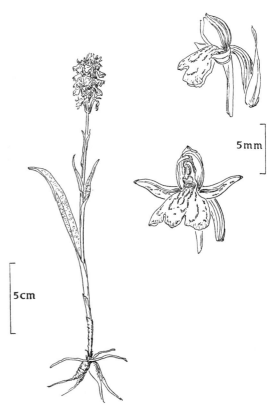

1630. *Dactylorchis incarnata* (L.) Vermeul.   Meadow
       Orchid   Purple, red or cream

1631. *Dactylorchis maculata* (L.) Vermeul.   Moorland
       Spotted Orchid   Pale pink with reddish dots

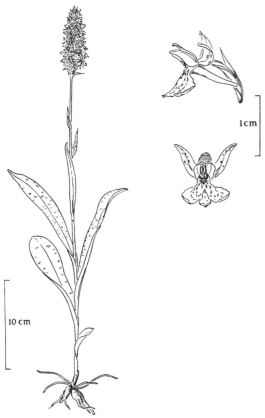

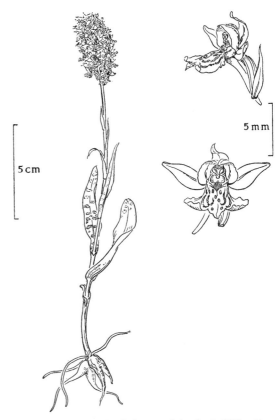

1632. *Dactylorchis fuchsii* (Druce) Vermeul.   Common
       Spotted Orchid   Pale pink with reddish lines

1633. *Dactylorchis fuchsii* ssp. *hebredensis* (Wilmott)
       H.-Harr.   Rosy magenta

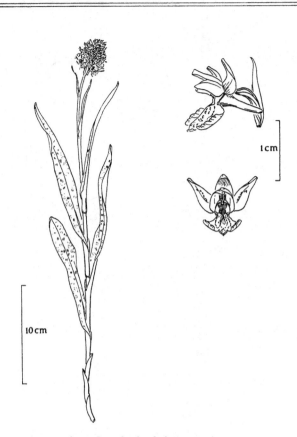

1634. *Dactylorchis fuchsii × praetermissa*

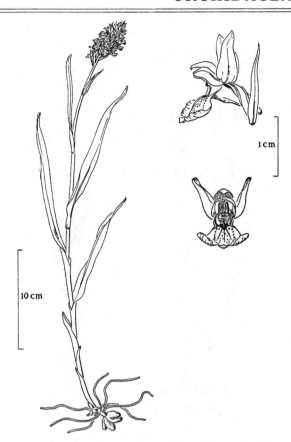

1635. *Dactylorchis praetermissa* (Druce) Vermeul.   Fen Orchid   Reddish-lilac

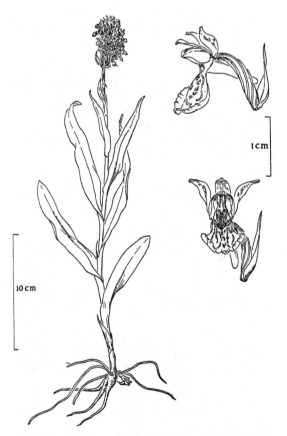

1636. *Dactylorchis purpurella* (T. & T. A. Steph.) Vermeul. Northern Fen Orchid   Deep purple

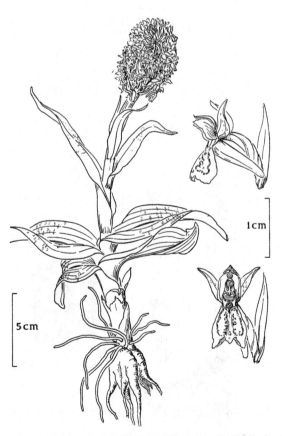

1637. *Dactylorchis majalis* (Rchb.) Vermeul.   Irish Marsh Orchid   Dark purple to pinkish mauve

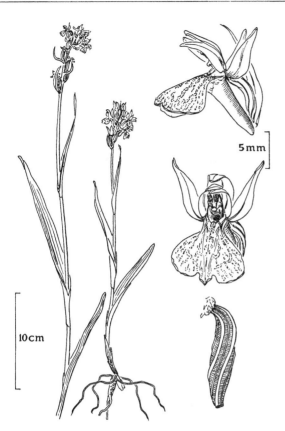

1638. *Dactylorchis traunsteineri* (Saut.) Vermeul   Narrow-
leaved Marsh Orchid   Reddish-purple

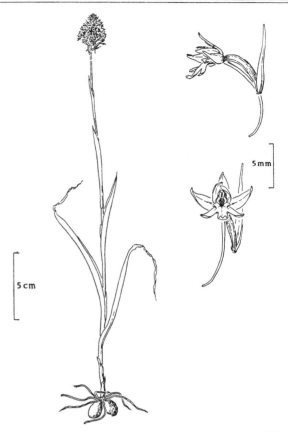

1639. *Anacamptis pyramidalis* (L.) Rich.   Pyramidal Orchid
Rosy purple

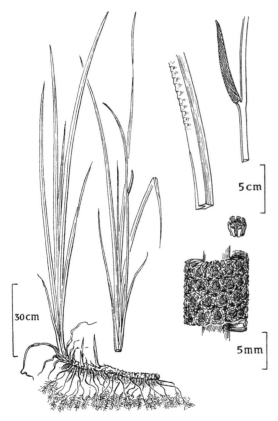

1640. *Acorus calamus* L.   Sweet Flag   Yellowish

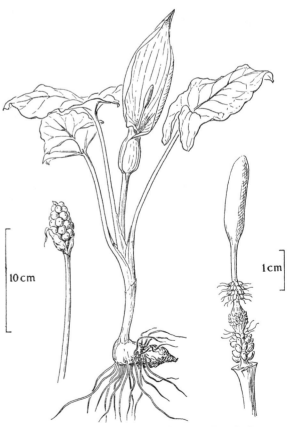

1641. *Arum maculatum* L.   Lords and Ladies, Cuckoo-
pint   Yellow-green

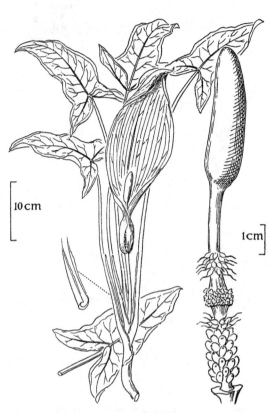

1642. *Arum italicum* Mill.   Pale green

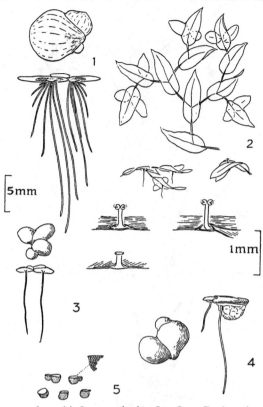

1643.   (1) *Lemna polyrrhiza* L.   Great Duckweed
         (2) *L. trisulca* L.   Ivy Duckweed
         (3) *L. minor* L.   Duckweed, Duck's-meat
         (4) *L. gibba* L.   Gibbous Duckweed
         (5) *Wolffia arrhiza* (L.) Hook. ex Wimm.

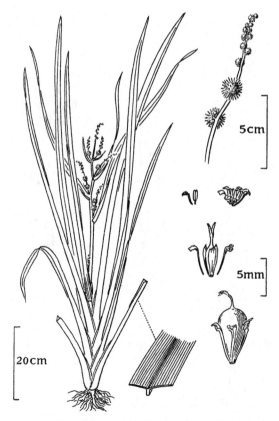

1644.   *Sparganium erectum* L.   Bur-reed   Greenish

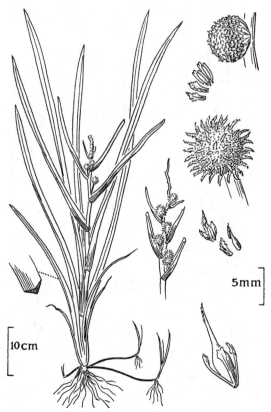

1645.   *Sparganium emersum* Rehm.   Unbranched Bur-reed
        Greenish

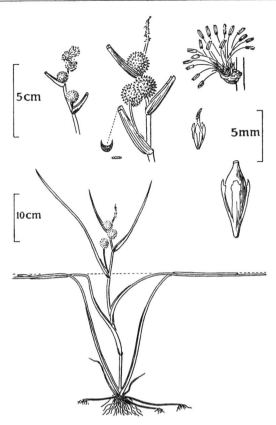

1646. *Sparganium angustifolium* Michx.   Floating Bur-reed
Greenish

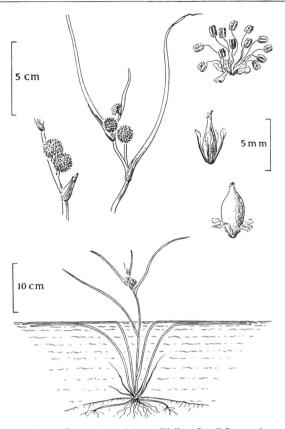

1647. *Sparganium minimum* Wallr.   Small Bur-reed
Greenish

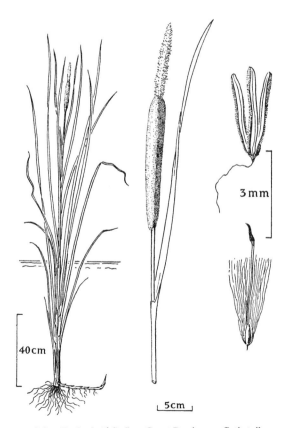

1648. *Typha latifolia* L.   Great Reedmace, Cat's-tail
Brown

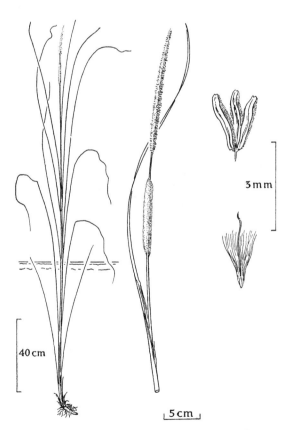

1649. *Typha angustifolia* L.   Lesser Reedmace   Brown

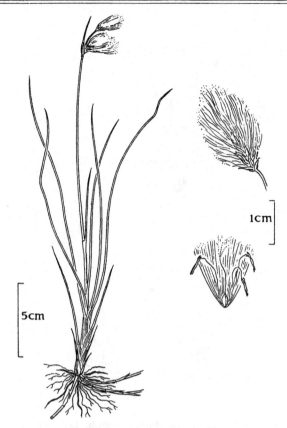

1650. *Eriophorum angustifolium* Honck.   Common
Cotton-grass   Brownish

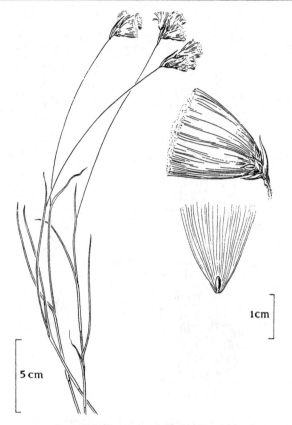

1651. *Eriophorum gracile* Roth _ Brownish

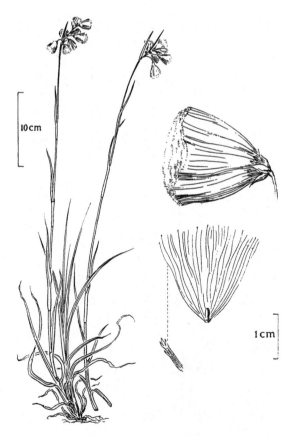

1652. *Eriophorum latifolium* Hoppe   Broad-leaved
Cotton-grass   Brownish

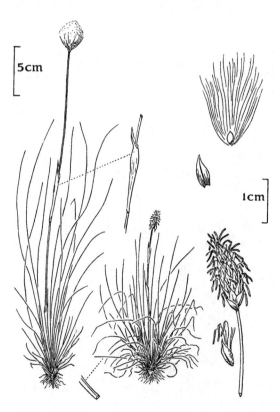

1653. *Eriophorum vaginatum* L.   Cotton-grass, Hare's-
tail   Brownish

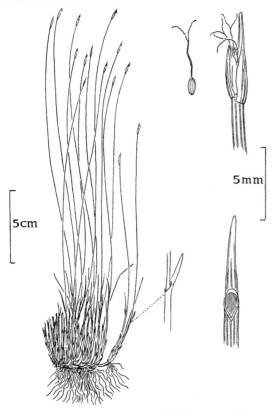

1654. *Trichophorum cespitosum* (L.) Hartmann Deer-
grass Brownish

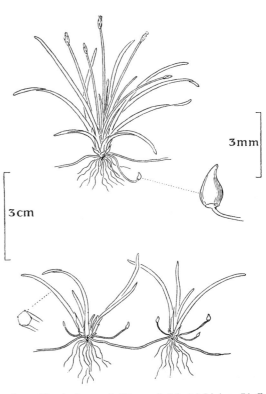

1655. *Eleocharis parvula* (Roem. & Schult.) Link ex Bluff,
Nees & Schau. Greenish

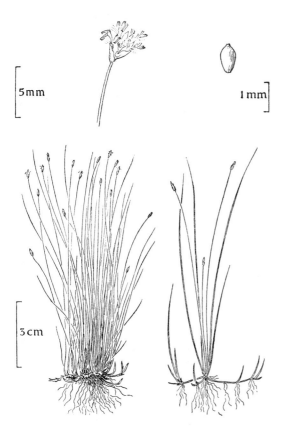

1656. *Eleocharis acicularis* (L.) Roem. & Schult. Slender
Spike-rush Brownish

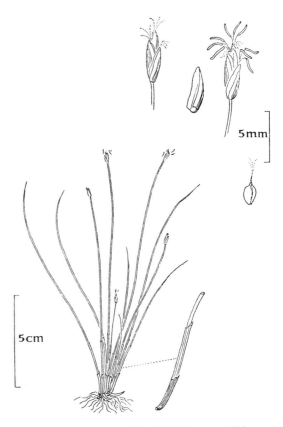

1657. *Eleocharis quinqueflora* (F. X. Hartmann) Schwarz
Few-flowered Spike-rush Brownish

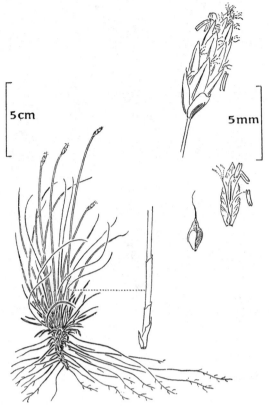

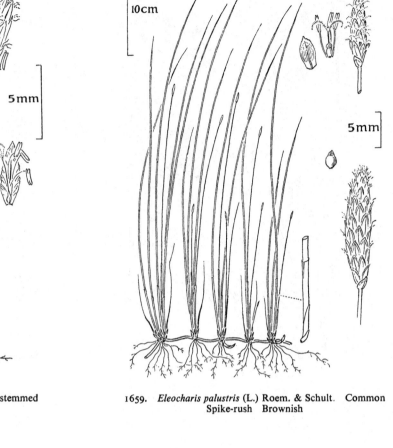

1658.   *Eleocharis multicaulis* (Sm.) Sm.   Many-stemmed
        Spike-rush   Brownish

1659.   *Eleocharis palustris* (L.) Roem. & Schult.   Common
        Spike-rush   Brownish

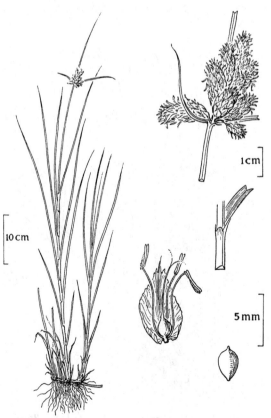

1660.   *Eleocharis uniglumis* (Link) Schult.   Brownish

1661.   *Scirpus maritimus* L.   Sea Club.

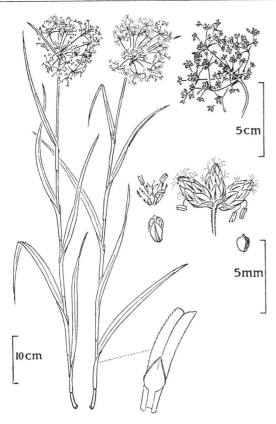

1662. *Scirpus sylvaticus* L. Wood Club-rush Greenish

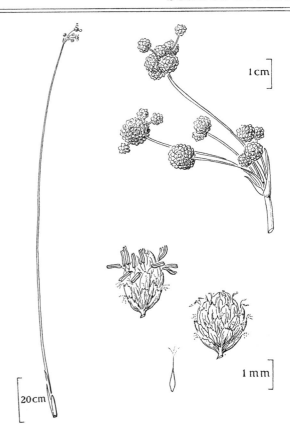

1663. *Holoschoenus vulgaris* Link Brownish

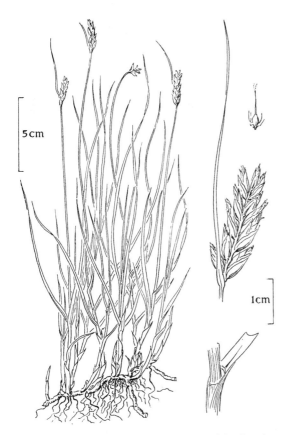

1664. *Blysmus compressus* (L.) Panz. ex Link Broad
Blysmus Reddish-brown

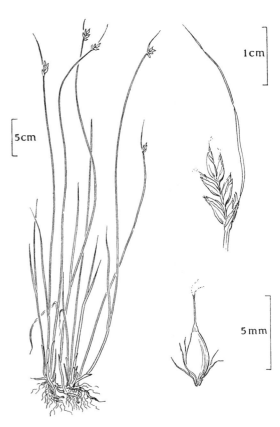

1665. *Blysmus rufus* (Huds.) Link Narrow Blysmus
Dark brown

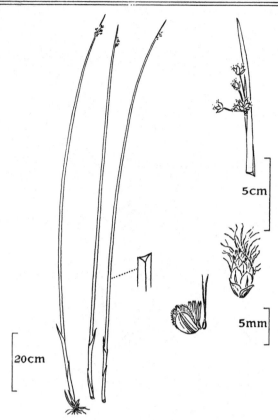

1666.   *Schoenoplectus triquetrus* (L.) Palla   Triangular
           Scirpus   Reddish-brown

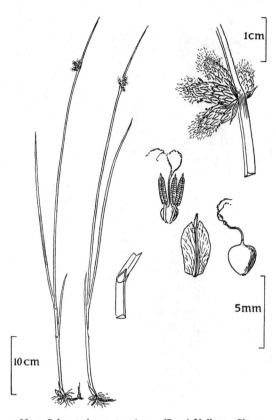

1667.   *Schoenoplectus americanus* (Pers.) Volkart   Sharp
           Scirpus   Brownish

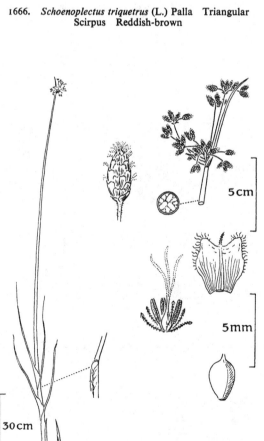

1668.   *Schoenoplectus lacustris* (L.) Palla   Bulrush   Brown

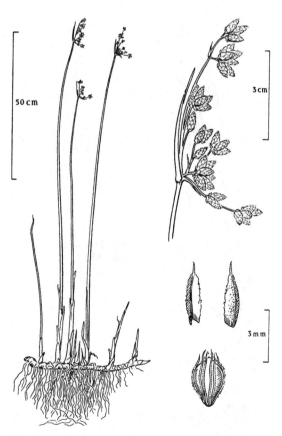

1669.   *Schoenoplectus tabernaemontani* (C. C. Gmel.)
           Palla   Glaucous Bulrush   Brown

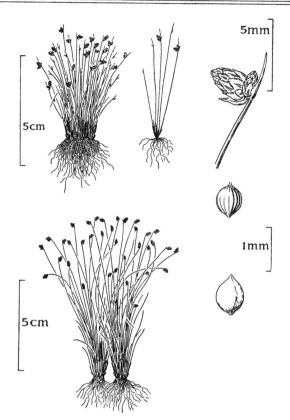

1670. *Isolepis setacea* (L.) R.Br. Bristle Scirpus Greenish (top) *I. cernua* (Vahl) Roem. & Schult. Nodding Scirpus Greenish (bottom)

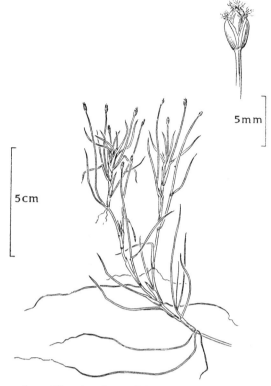

1671. *Eleogeiton fluitans* (L.) Link Floating Scirpus Greenish

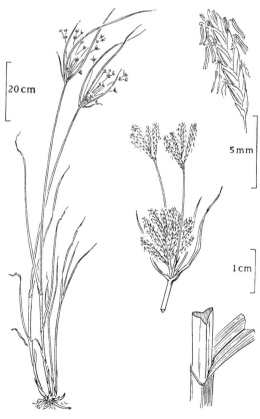

1672. *Cyperus longus* L. Galingale Reddish-brown

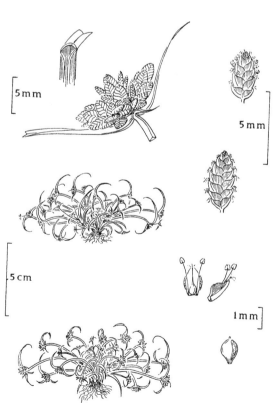

1673. *Cyperus fuscus* L. Brown Cyperus Reddish-brown

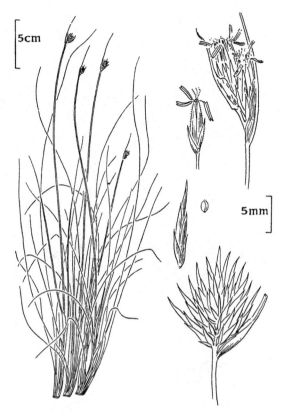

1674. *Schoenus nigricans* L. Bog-rush Blackish

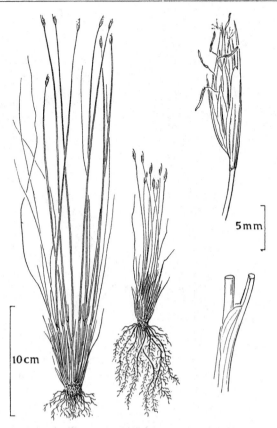

1675. *Schoenus ferrugineus* L. Dark reddish-brown

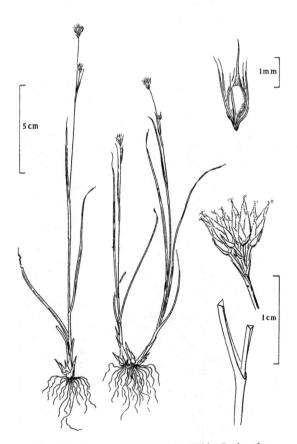

1676. *Rhynchospora alba* (L.) Vahl White Beak-sedge
Whitish

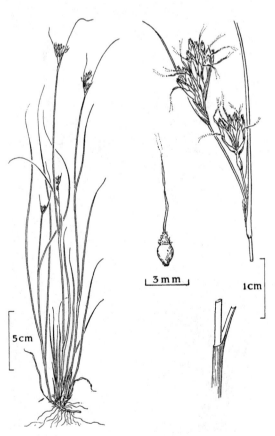

1677. *Rhynchospora fusca* (L.) Ait. f. Brown Beak-sedge
Reddish-brown

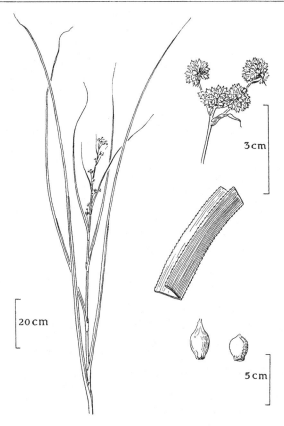

1678. *Cladium mariscus* (L.) Pohl Sedge Reddish-brown

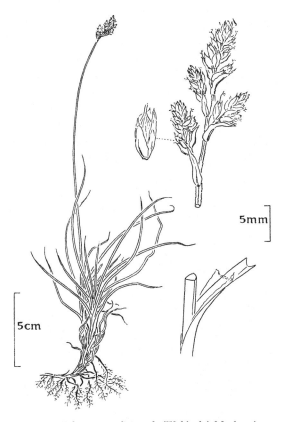

1679. *Kobresia simpliciuscula* (Wahlenb.) Mackenzie
Brown

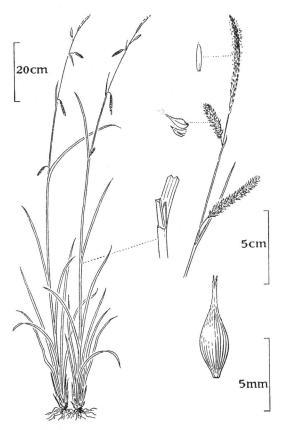

1680. *Carex laevigata* Sm. Smooth Sedge Brownish

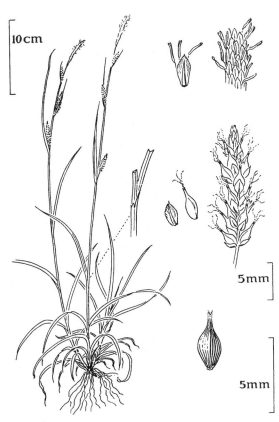

1681. *Carex distans* L. Distant Sedge Greenish

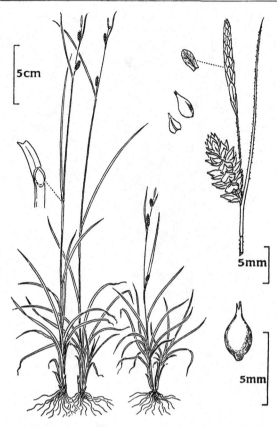

1682.   *Carex punctata* Gaud.   Dotted Sedge   Brownish

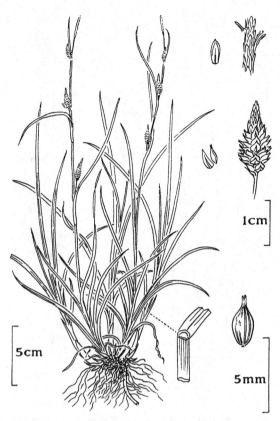

1683.   *Carex hostiana* DC.   Tawny Sedge   Brown

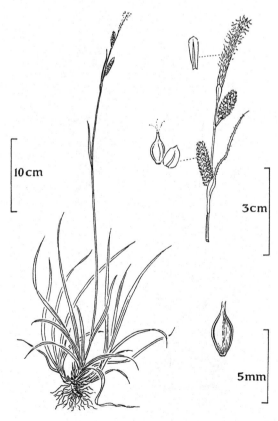

1684.   *Carex binervis* Sm.   Ribbed Sedge   Dark brown

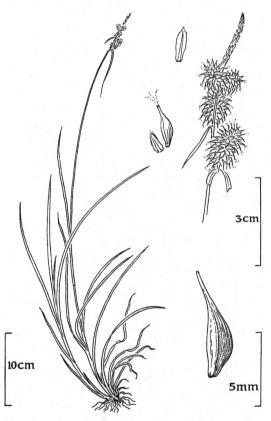

1685.   *Carex flava* L.   Greenish

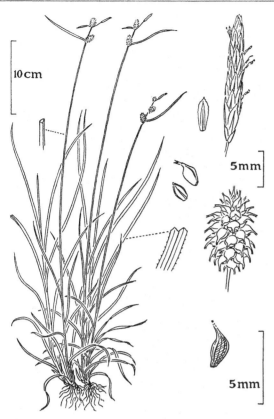

1686.   *Carex lepidocarpa* Tausch   Yellow Sedge   Greenish

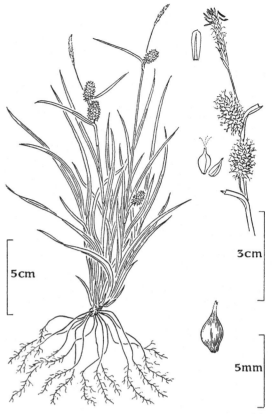

1687.   *Carex demissa* Hornem.   Yellow Sedge   Greenish

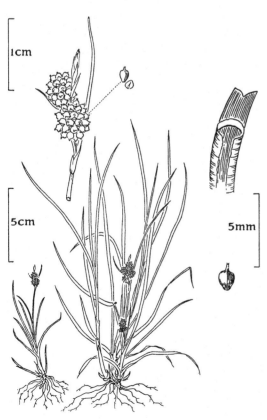

1688.   *Carex serotina* Merát   Yellow-brown

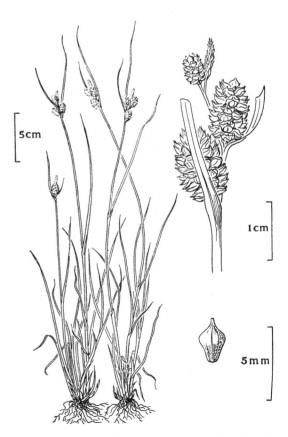

1689.   *Carex extensa* Good.   Long-bracted Sedge   Greenish

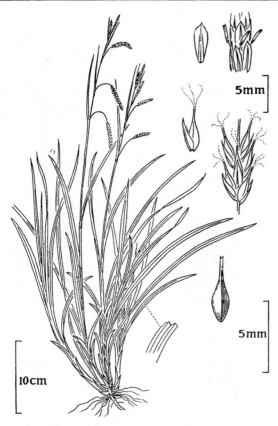

1690.   *Carex sylvatica* Huds.   Wood Sedge   Greenish

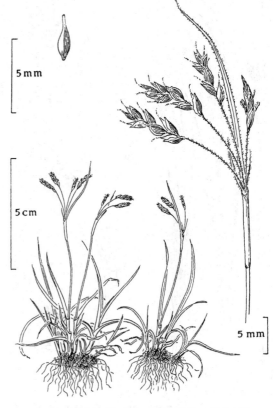

1691.   *Carex capillaris* L.   Hair Sedge   Brownish

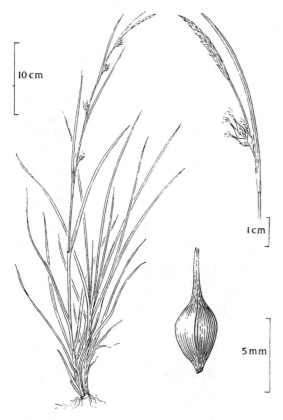

1692.   *Carex depauperata* Curt. ex With.   Brownish

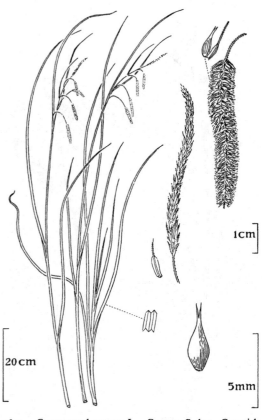

1693.   *Carex pseudocyperus* L.   Cyperus Sedge   Greenish

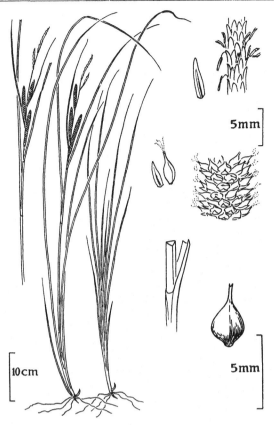

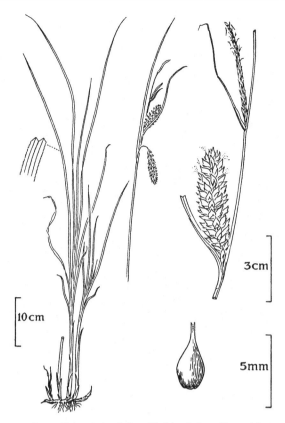

1694. *Carex rostrata* Stokes   Beaked Sedge, Bottle Sedge
Brownish

1695. *Carex vesicaria* L.   Bladder Sedge   Brownish

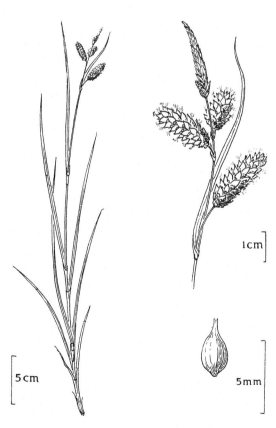

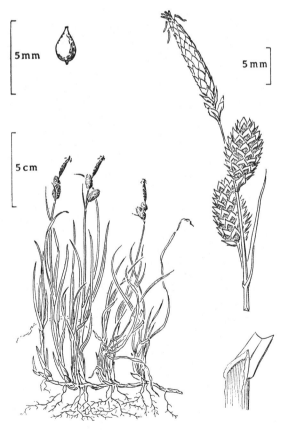

1696. *Carex stenolepis* Less.   Dark brown

1697. *Carex saxatilis* L.   Russet Sedge   Dark brown

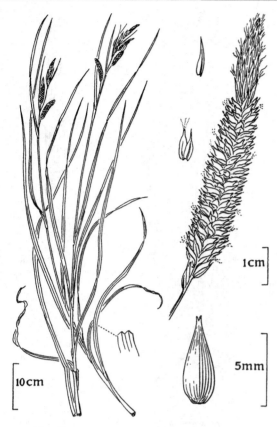

1698.   *Carex riparia* L.   Great Pond-sedge   Brown

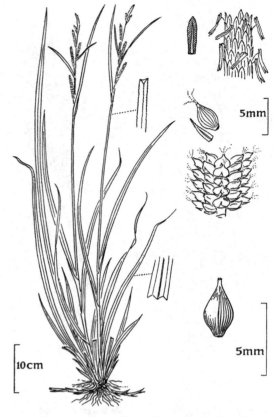

1699.   *Carex acutiformis* Ehrh.   Lesser Pond-sedge   Brown

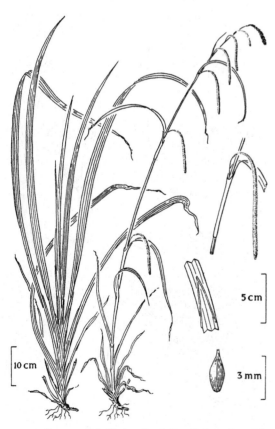

1700.   *Carex pendula* Huds.   Pendulous Sedge   Greenish

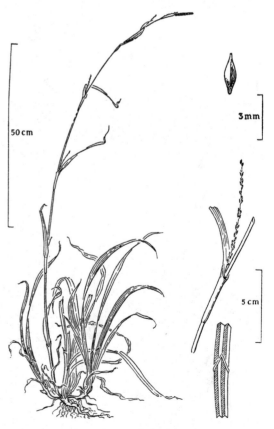

1701.   *Carex strigosa* Huds.   Greenish

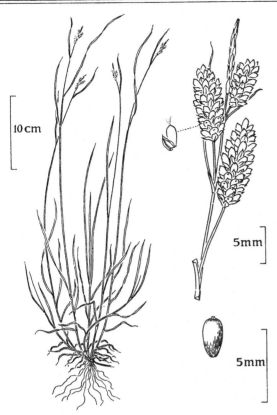

1702. *Carex pallescens* L. Pale Sedge Pale green

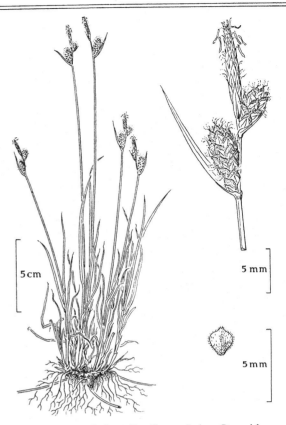

1703. *Carex filiformis* L. Downy Sedge Brownish

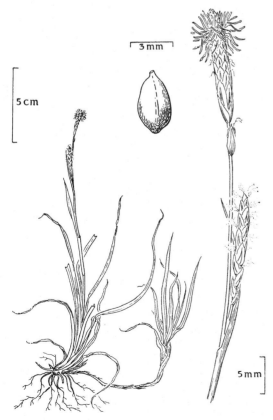

1704. *Carex panicea* L. Carnation-grass Brownish

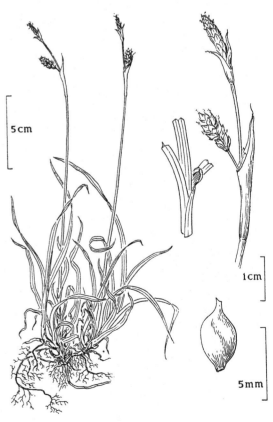

1705. *Carex vaginata* Tausch Brownish

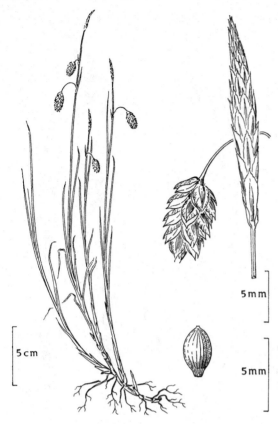

1706.   *Carex limosa* L.   Mud Sedge   Brown

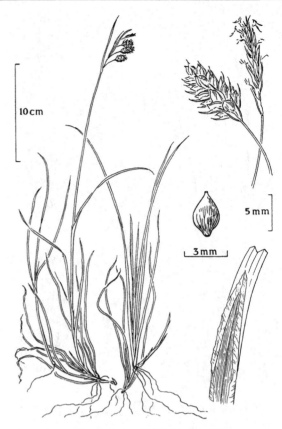

1707.   *Carex paupercula* Michx.   Dark brown

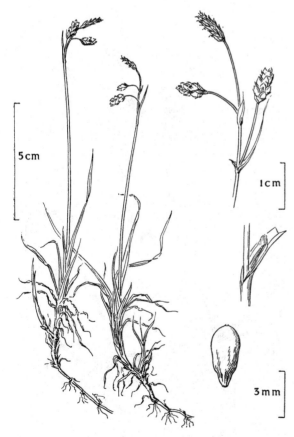

1708.   *Carex rariflora* (Wahlenb.) Sm.   Brownish

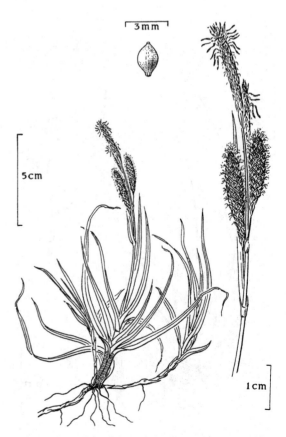

1709.   *Carex flacca* Schreb.   Carnation-grass   Blackish

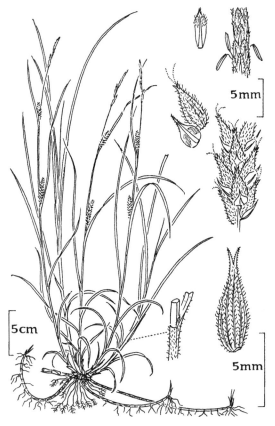

1710. *Carex hirta* L. Hammer Sedge Greenish

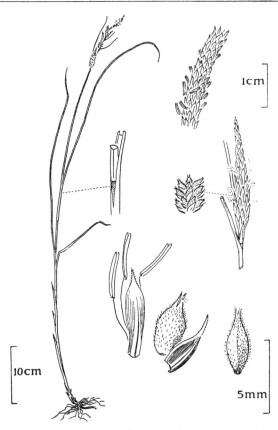

1711. *Carex lasiocarpa* Ehrh. Slender Sedge Brownish

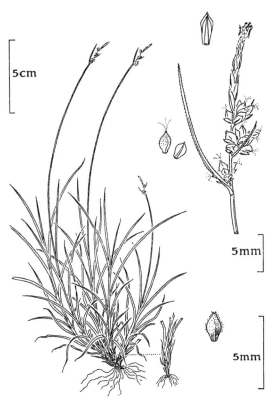

1712. *Carex pilulifera* L. Pill-headed Sedge Brownish

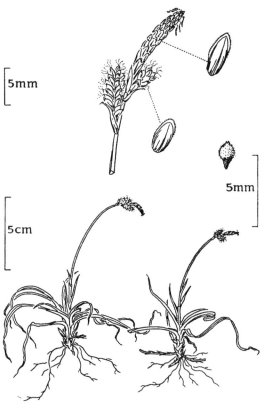

1713. *Carex ericetorum* Poll. Brown

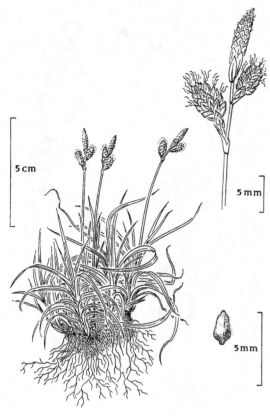

1714. *Carex caryophyllea* Latour.   Spring Sedge   Brown

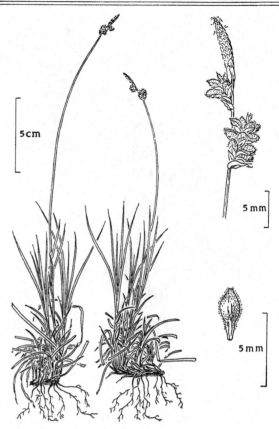

1715. *Carex montana* L.   Mountain Sedge   Brown

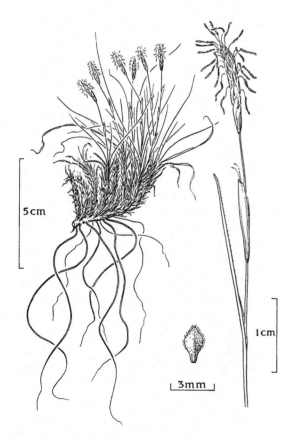

1716. *Carex humilis* Leyss.   Dwarf Sedge   Brownish

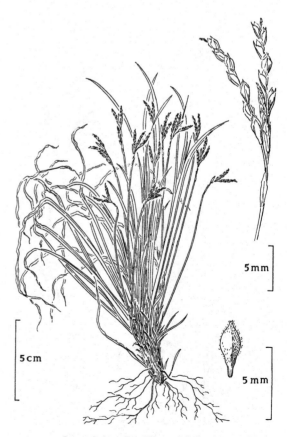

1717. *Carex digitata* L.   Fingered Sedge   Brownish

5-2

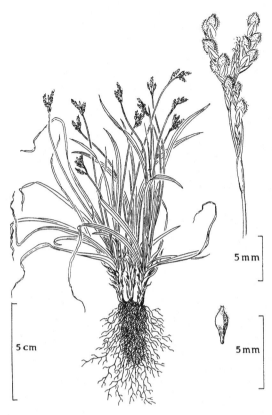

1718. *Carex ornithopoda* Willd.   Bird's-foot Sedge
Brownish

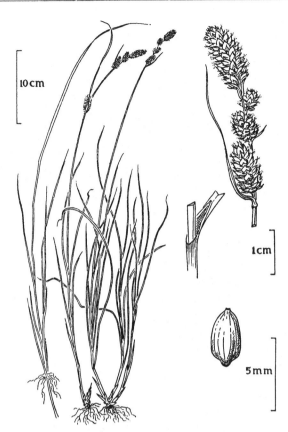

1719.   *Carex buxbaumii* Wahlenb.   Dark brown

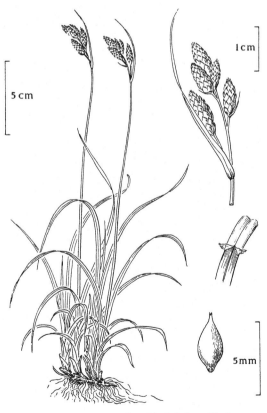

1720.   *Carex atrata* L.   Black Sedge   Black

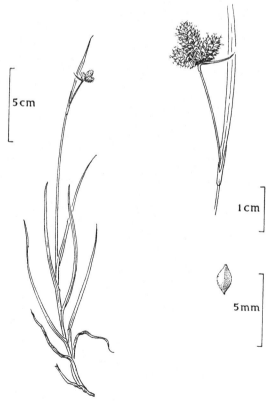

1721.   *Carex norvegica* Retz.   Black

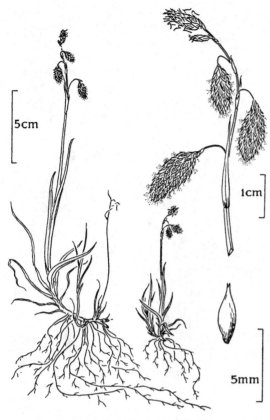

1722. *Carex atrofusca* Schkuhr   Black

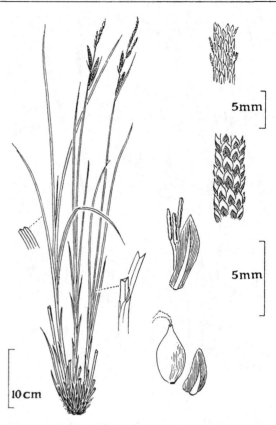

1723. *Carex elata* All.   Tufted Sedge   Dark brown

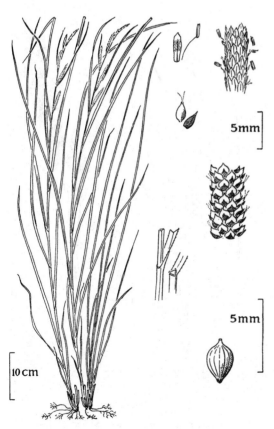

1724. *Carex acuta* L.   Tufted Sedge   Dark brown

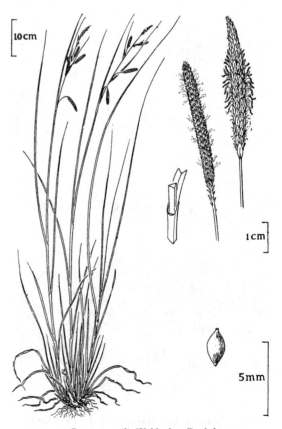

1725. *Carex aquatilis* Wahlenb.   Dark brown

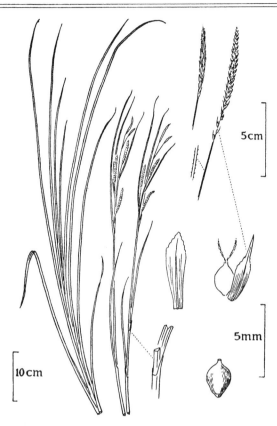

1726. *Carex recta* Boott Dark brown

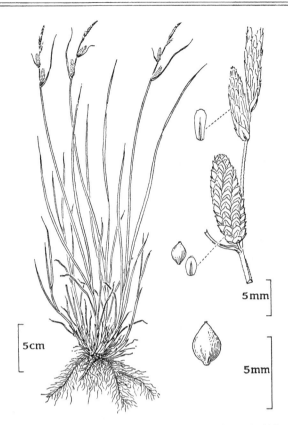

1727. *Carex nigra* (L.) Reichard Common Sedge Blackish

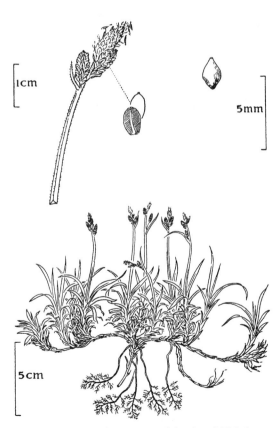

1728. *Carex bigelowii* Torr. ex Schwein. Stiff Sedge
Blackish

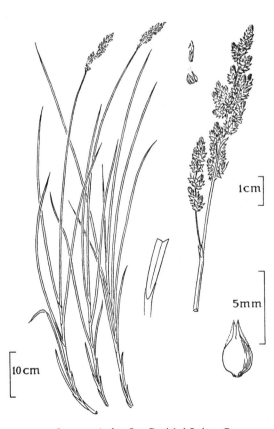

1729. *Carex paniculata* L. Panicled Sedge Brown

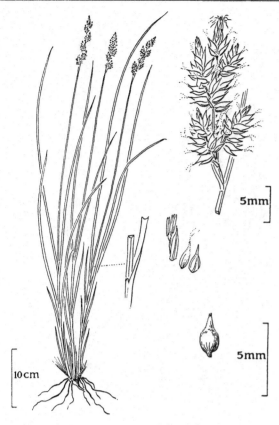

1730. *Carex appropinquata* Schumacher   Brown

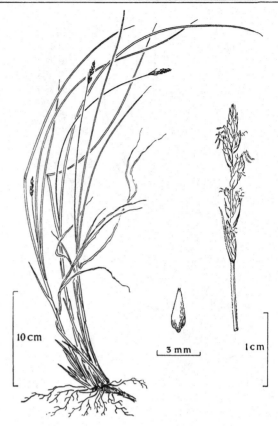

1731. *Carex diandra* Schrank   Brownish

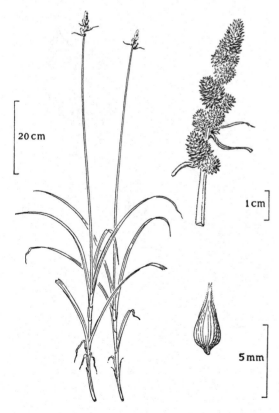

1732. *Carex otrubae* Podp.   False Fox-sedge   Greenish

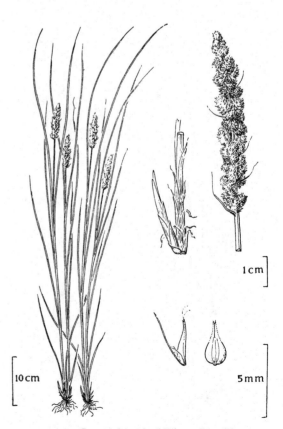

1733. *Carex vulpinoidea* Michx.   Greenish

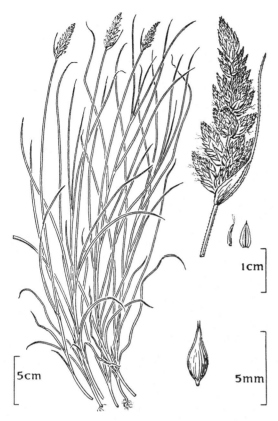

1734.   *Carex disticha* Huds.   Brown Sedge   Brownish

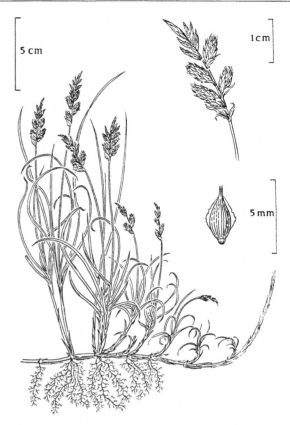

1735.   *Carex arenaria* L.   Sand Sedge   Brownish

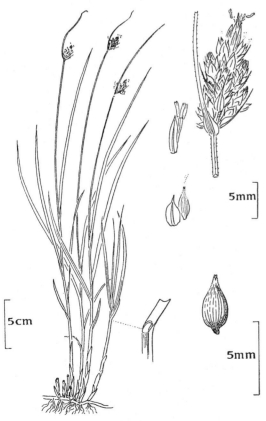

1736.   *Carex divisa* Huds.   Divided Sedge   Brownish

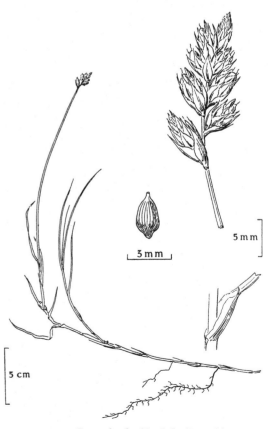

1737.   *Carex chordorrhiza* L.f.   Brownish

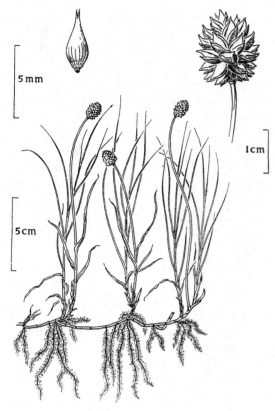

1738. *Carex maritima* Gunn. Curved Sedge Brown

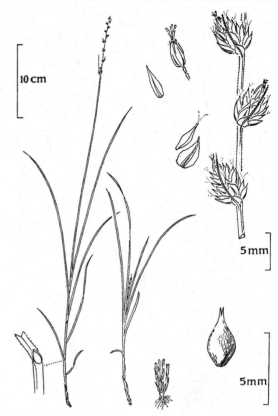

1739. *Carex divulsa* Stokes Grey Sedge Greenish

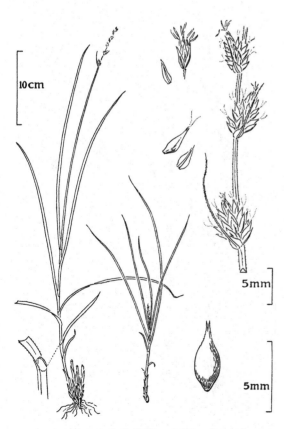

1740. *Carex polyphylla* Kar. & Kir. Greenish

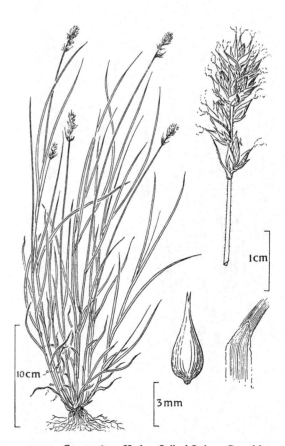

1741. *Carex spicata* Huds. Spiked Sedge Greenish

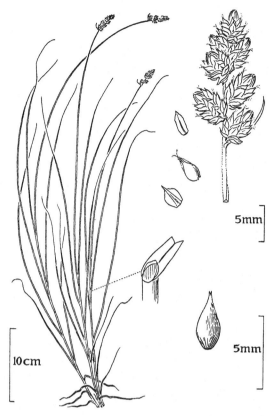

1742. *Carex muricata* L. Prickly Sedge   Greenish

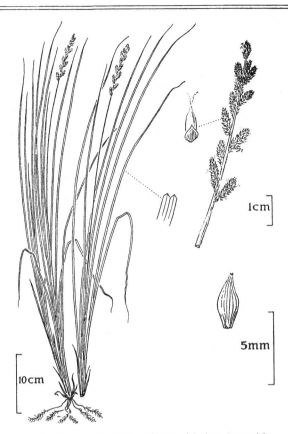

1743. *Carex elongata* L. Elongated Sedge   Brownish

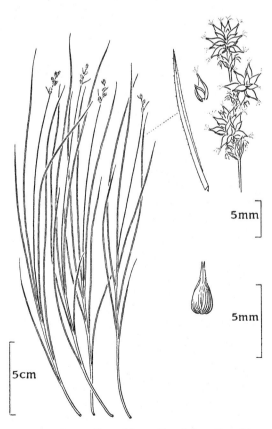

1744. *Carex echinata* Murr. Star Sedge   Greenish

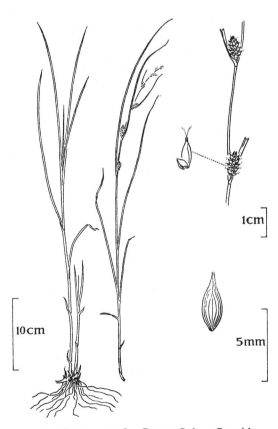

1745. *Carex remota* L. Remote Sedge   Greenish

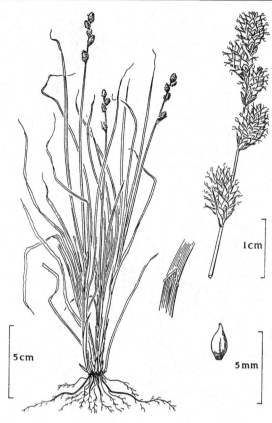

1746. *Carex curta* Good.   White Sedge   Whitish

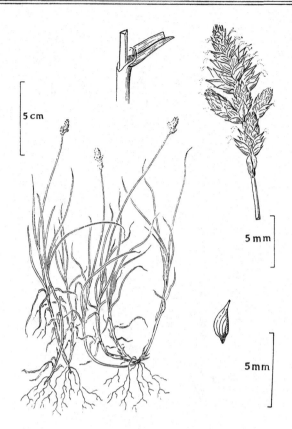

1747. *Carex lachenalii* Schkuhr   Brownish

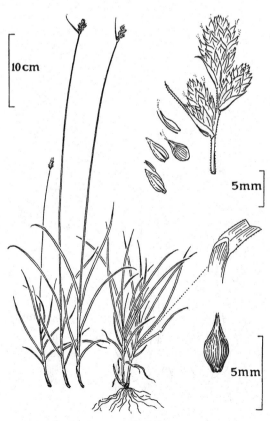

1748. *Carex ovalis* Good.   Oval Sedge   Brownish

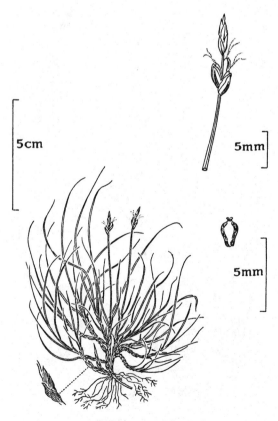

1749. *Carex rupestris* All.   Brown

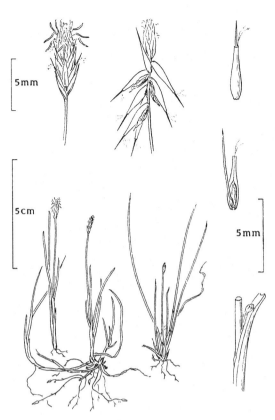

1750.   *Carex microglochin* Wahlenb.   Brownish

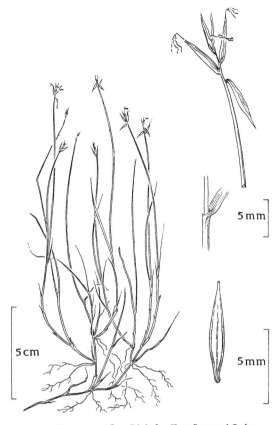

1751.   *Carex pauciflora* Lightf.   Few-flowered Sedge
Brownish

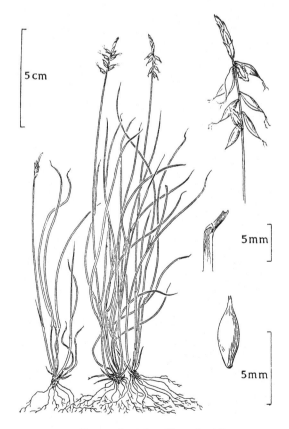

1752.   *Carex pulicaris* L.   Flea-sedge   Brown

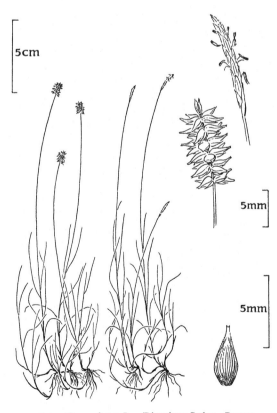

1753.   *Carex dioica* L.   Dioecious Sedge   Brown

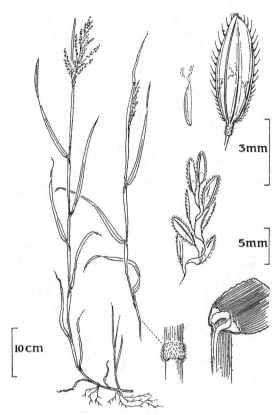

1754.   *Leersia oryzoides* (L.) Sw.   Cut-grass   Green

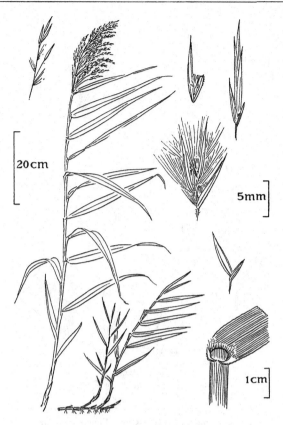

1755.   *Phragmites communis* Trin.   Reed   Purple

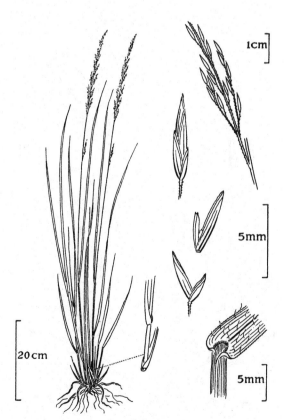

1756.   *Molinia caerulea* (L.) Moench   Purple Moor-grass
Green or purplish

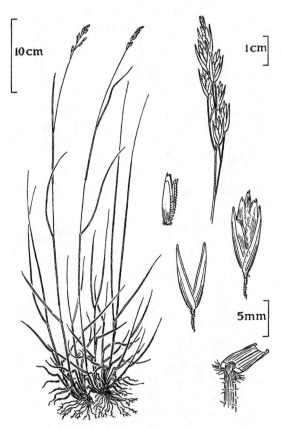

1757.   *Sieglingia decumbens* (L.) Bernh.   Heath Grass
Green

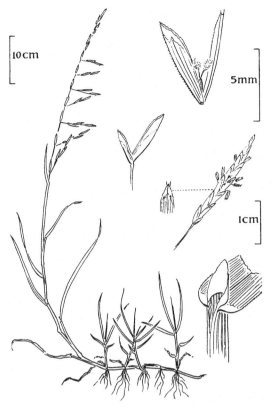

1758.   *Glyceria fluitans* (L.) R.Br.   Flote-grass   Green

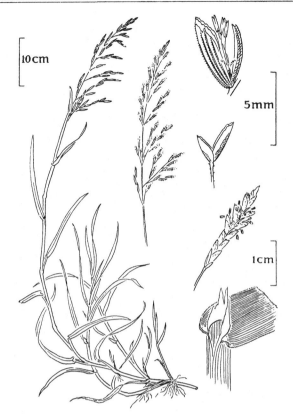

1759.   *Glyceria plicata* Fr.   Green

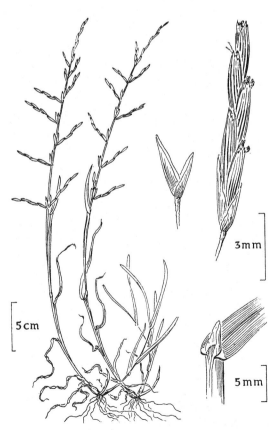

1760.   *Glyceria declinata* Bréb.   Green

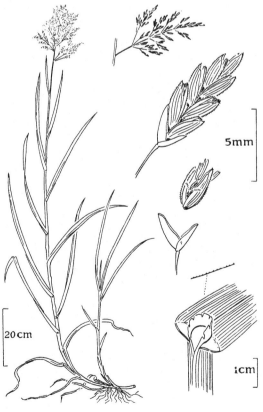

1761.   *Glyceria maxima* (Hartm.) Holmberg   Reed-grass
Green

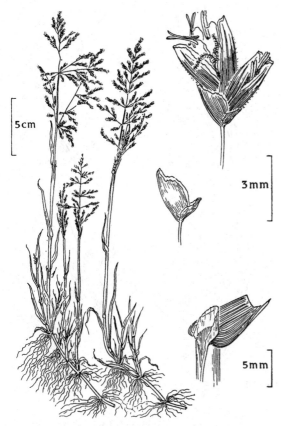

1762. *Catabrosa aquatica* (L.) Beauv. Water Whorl-grass
Green or purplish

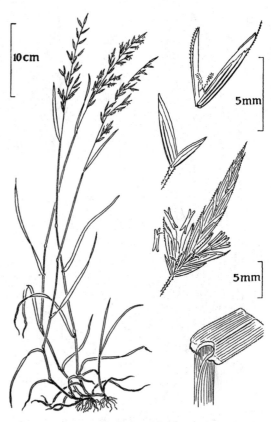

1763. *Festuca pratensis* Huds. Meadow Fescue Green
or purplish

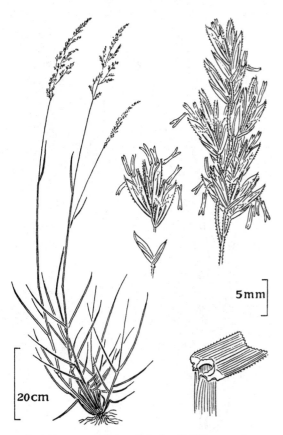

1764. *Festuca arundinacea* Schreb. Tall Fescue Green
or purplish

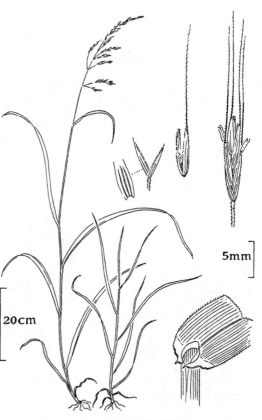

1765. *Festuca gigantea* (L.) Vill. Tall Brome Green

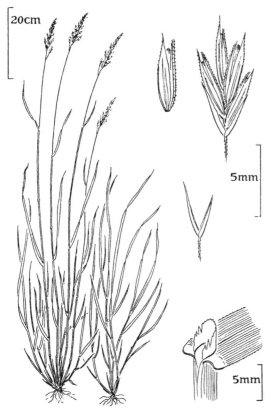

1766.   *Festuca altissima* All.   Green

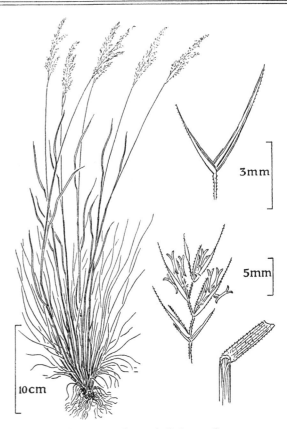

1767.   *Festuca heterophylla* Lam.   Green

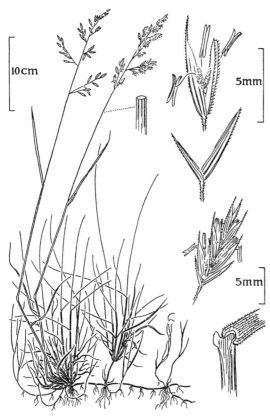

1768.   *Festuca rubra* L.   Creeping Fescue   Reddish, purplish or green

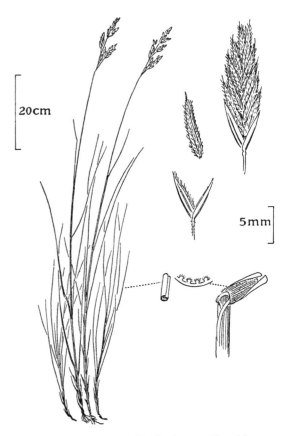

1769.   *Festuca juncifolia* St.-Amans   Greenish

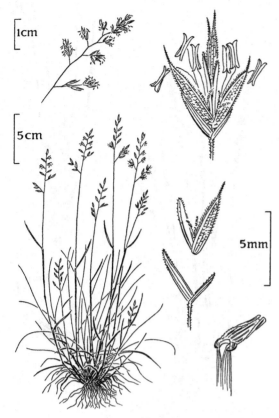

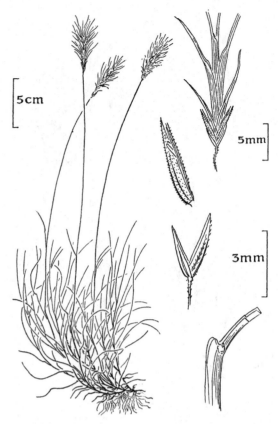

1770.   *Festuca ovina* L.   Sheep's Fescue   Greenish or purplish

1771.   *Festuca vivipara* (L.) Sm.   Greenish

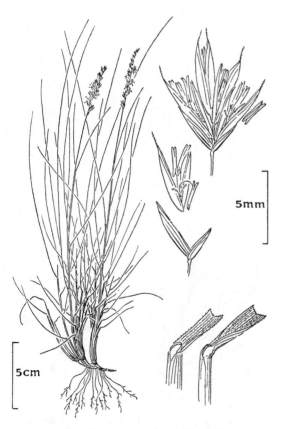

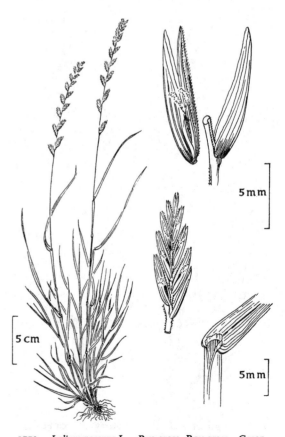

1772.   *Festuca glauca* Lam.   Greenish

1773.   *Lolium perenne* L.   Rye-grass, Ray-grass   Green

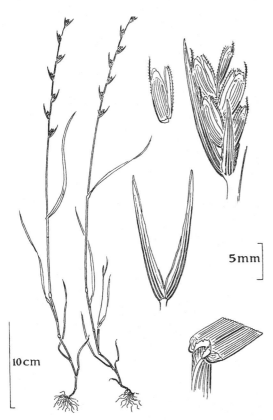

1774.   *Lolium temulentum* L.   Darnel   Green

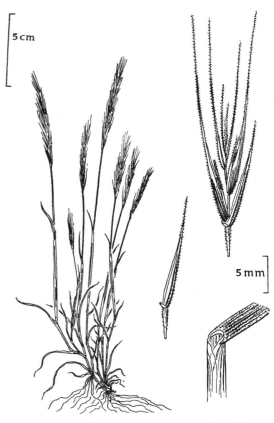

1775.   *Vulpia membranacea* (L.) Dum.   Greenish

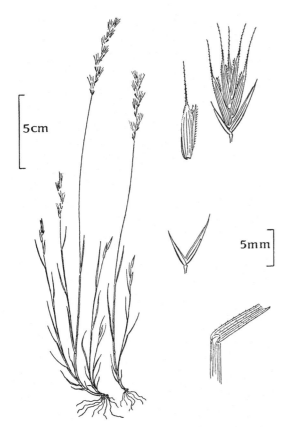

1776.   *Vulpia bromoides* (L.) S. F. Gray   Barren Fescue
Greenish

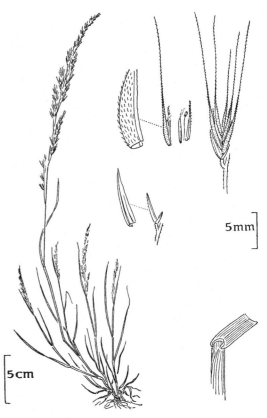

1777.   *Vulpia myuros* (L.) C. C. Gmel.   Rat's-tail Fescue
Greenish

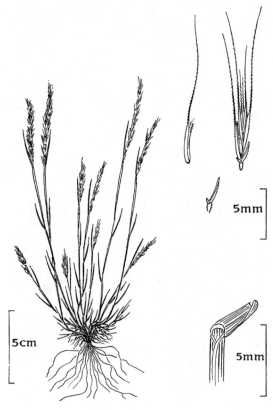

1778. *Vulpia ambigua* (Le Gall) More   Greenish or purplish

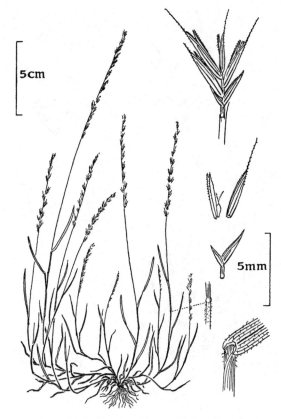

1779. *Nardurus maritimus* (L.) Murb.   Greenish

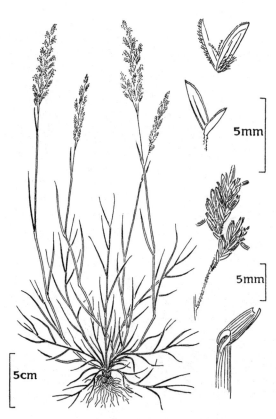

1780. *Puccinellia maritima* (Huds.) Parl.   Sea Poa
Greenish

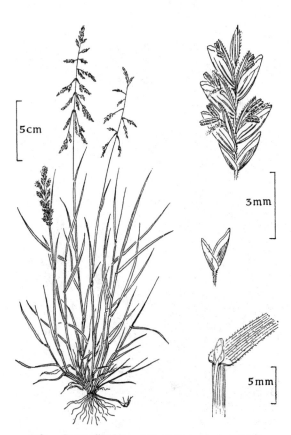

1781. *Puccinellia distans* (Jacq.) Parl.   Reflexed Poa
Greenish

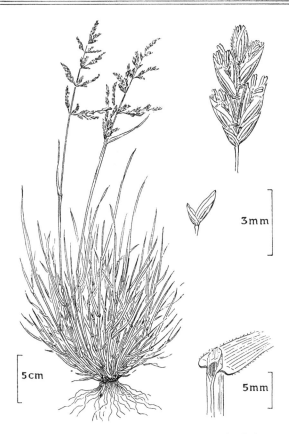

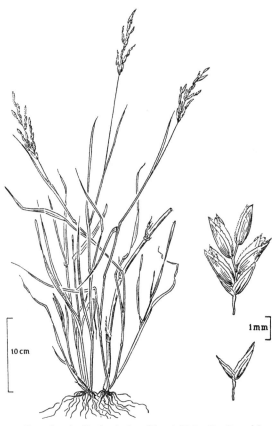

1782.   *Puccinellia pseudodistans* (Crép.) Jans. & Wacht.
Greenish

1783.   *Puccinellia fasciculata* (Torr.) Bicknell   Greenish

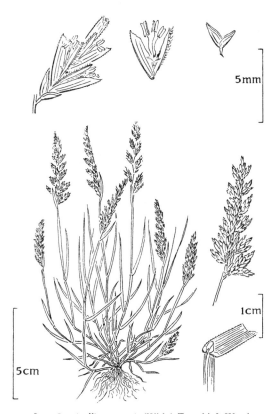

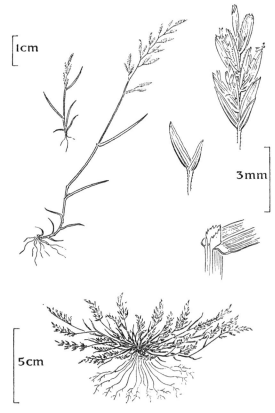

1784.   *Puccinellia rupestris* (With.) Fernald & Weath.
Procumbent Poa   Greenish

1785.   *Catapodium rigidum* (L.) C. E. Hubbard   Hard Poa
Greenish

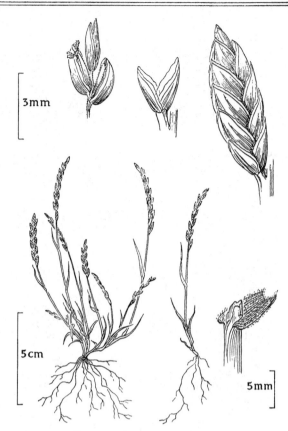

1786. *Catapodium marinum* (L.) C. E. Hubbard　Darnel
Poa　Greenish

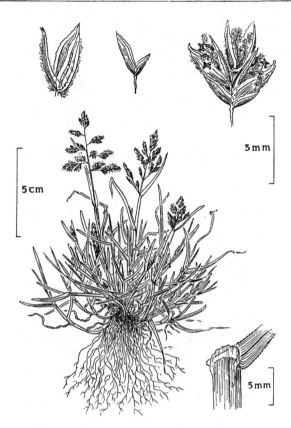

1787. *Poa annua* L.　Annual Poa　Green or purplish

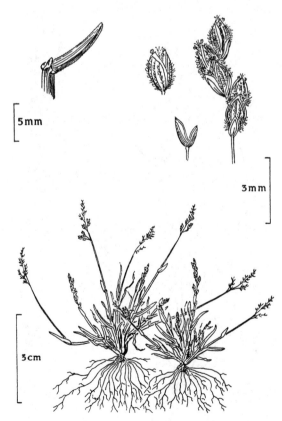

1788. *Poa infirma* Kunth　Green

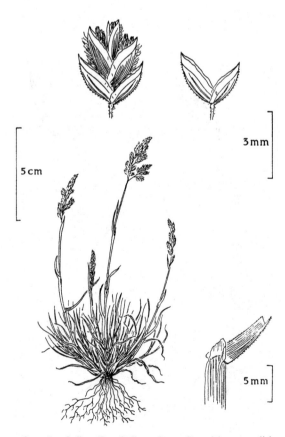

1789. *Poa bulbosa* L.　Bulbous Poa　Greenish or purplish

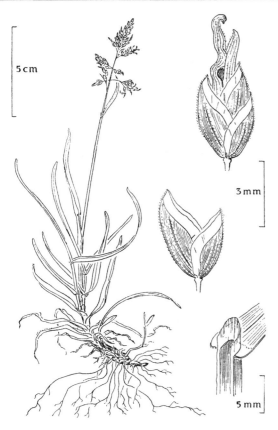

1790. *Poa alpina* L. Alpine Poa Greenish or purplish

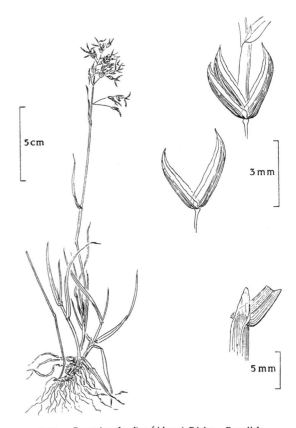

1791. *Poa × jemtlandica* (Almq.) Richt. Purplish

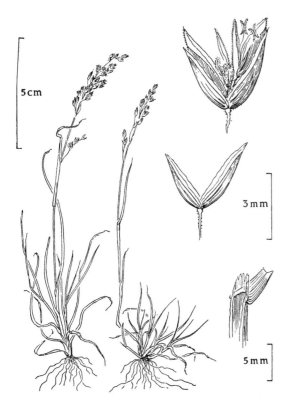

1792. *Poa flexuosa* Sm. Wavy Poa Green or purplish

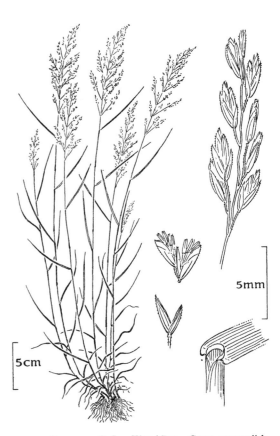

1793. *Poa nemoralis* L. Wood Poa Green or purplish

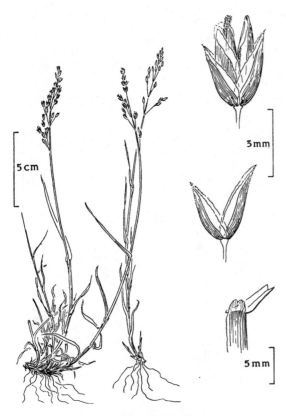

1794. *Poa glauca* Vahl   Glaucous

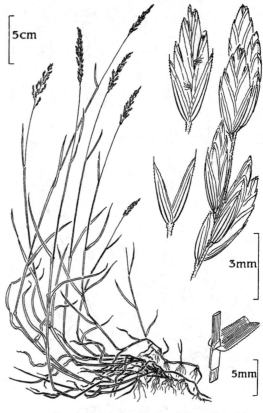

1795. *Poa compressa* L.   Flattened Poa   Greenish

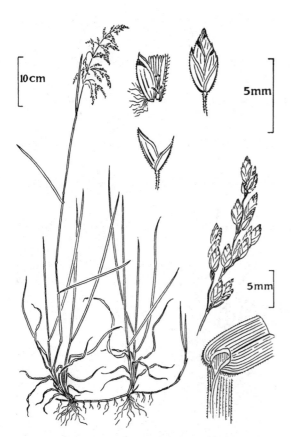

1796. *Poa pratensis* L. ssp. *pratensis*   Meadow-grass
Green or purplish

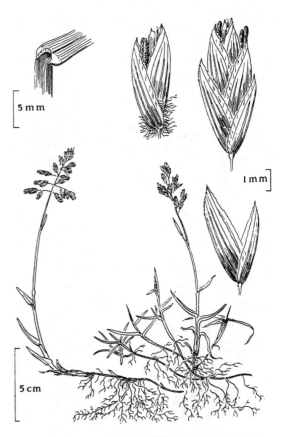

1797. *Poa pratensis* ssp. *irrigata* (Lindm.) Lindb.f.   Green
or purplish

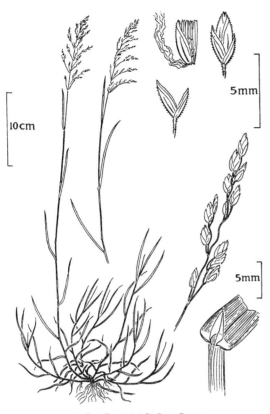

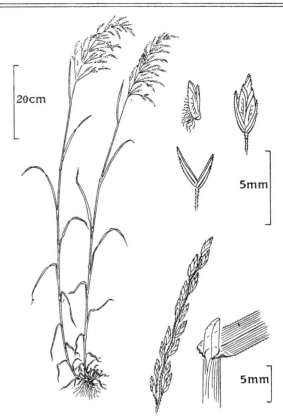

1798.   *Poa trivialis* L.   Green

1799.   *Poa palustris* L.   Green

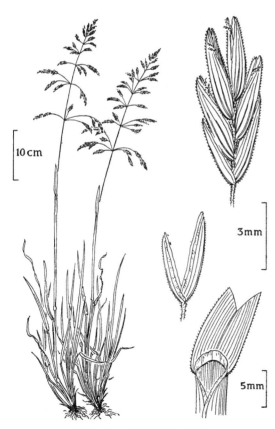

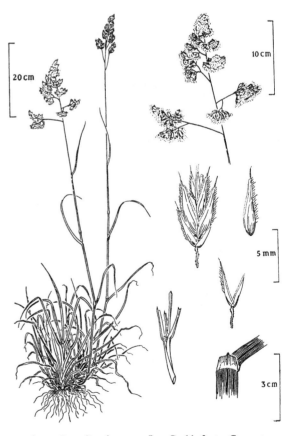

1800.   *Poa chaixii* Vill.   Green

1801.   *Dactylis glomerata* L.   Cock's-foot   Green or
purplish

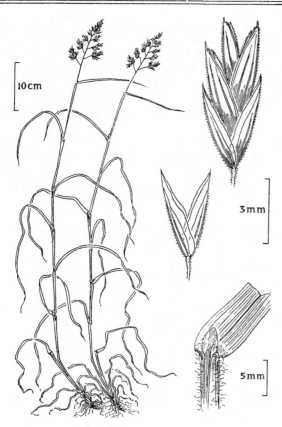

1802. *Dactylis polygama* Horvat. Green

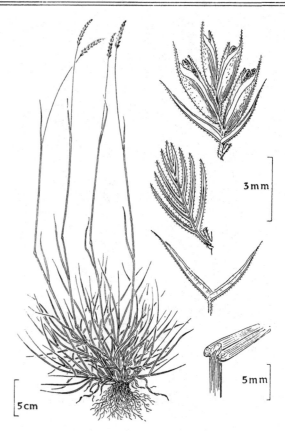

1803. *Cynosurus cristatus* L. Crested Dog's-tail Green
or purplish

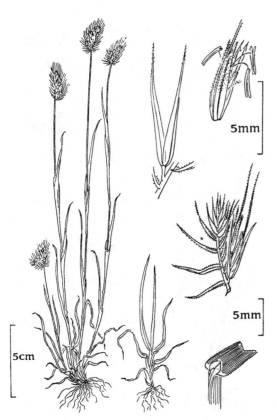

1804. *Cynosurus echinatus* L. Green

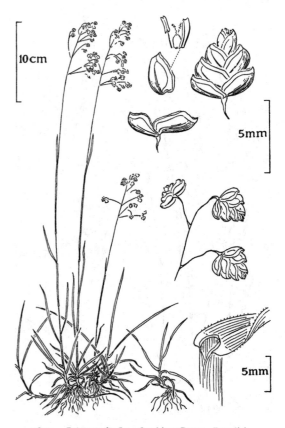

1805. *Briza media* L. Quaking Grass Purplish

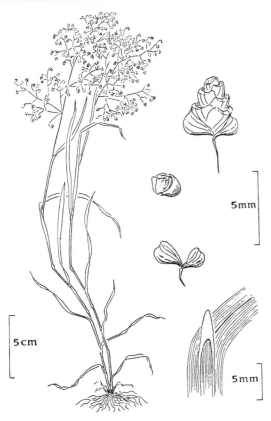

1806.   *Briza minor* L.   Green

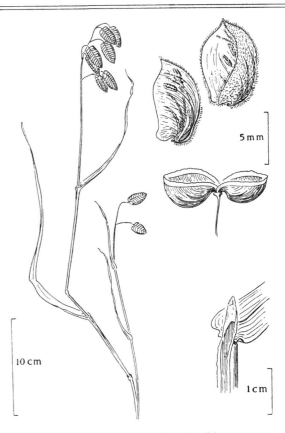

1807.   *Briza maxima* L.   Purplish

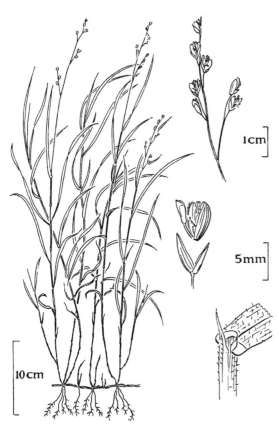

1808.   *Melica uniflora* Retz.   Wood Melick   Purplish

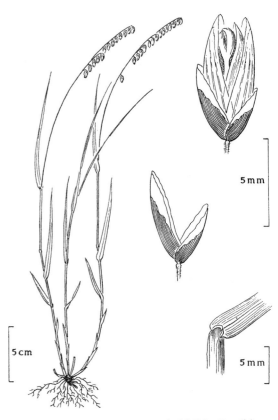

1809.   *Melica nutans* L.   Mountain Melick   Purplish

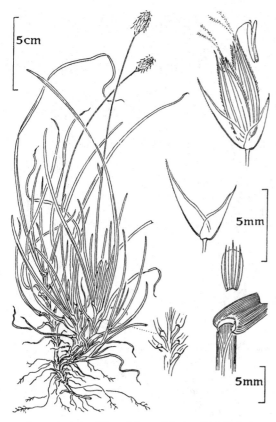

1810.   *Sesleria albicans* Kit. ex Schult.   Blue Sesleria
Blue-grey

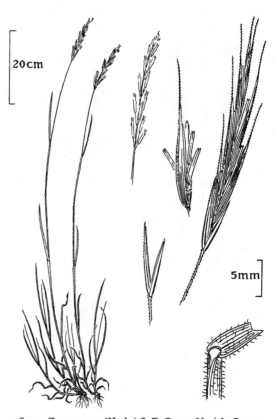

1811.   *Zerna erecta* (Huds.) S. F. Gray   Upright Brome
Purplish

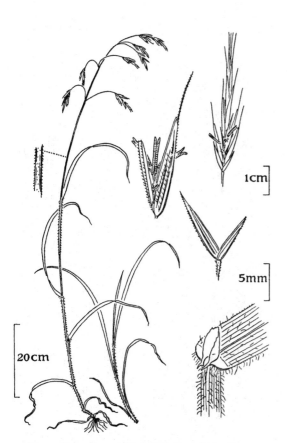

1812.   *Zerna ramosa* (Huds.) Lindm.   Hairy Brome
Purplish or glaucous

1813.   *Anisantha sterilis* (L.) Nevski   Barren Brome
Green or purplish

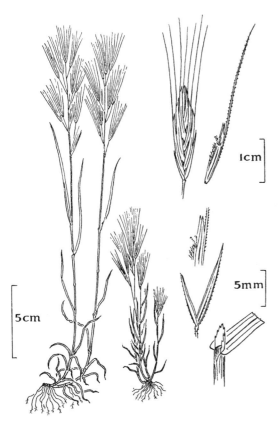

1814. *Anisantha madritensis* (L.) Nevski   Compact Brome
Purplish

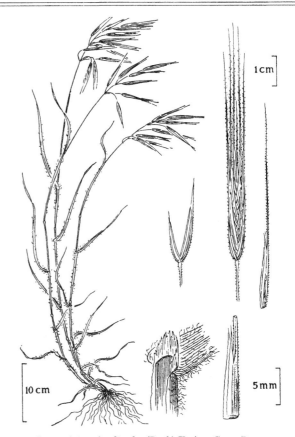

1815. *Anisantha diandra* (Roth) Tutin   Great Brome
Greenish

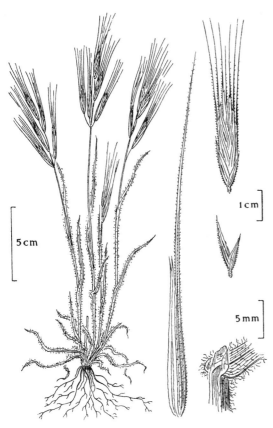

1816. *Anisantha rigida* (Roth) Hyl.   Greenish

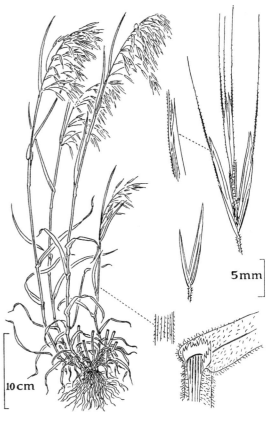

1817. *Anisantha tectorum* (L.) Nevski   Greenish

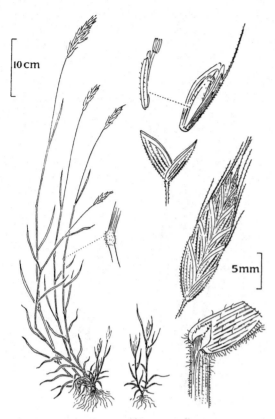

1818. *Bromus mollis* L. Lop-grass Green

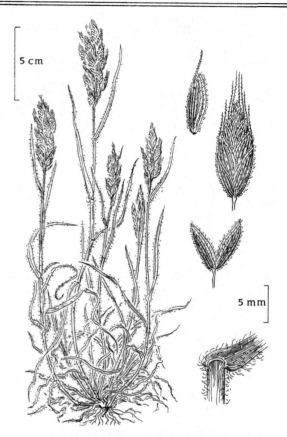

1819. *Bromus ferronii* Mabille Green

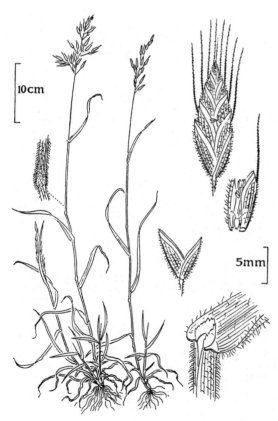

1820. *Bromus thominii* Hard. Lop-grass Green

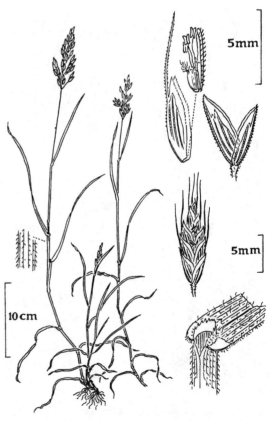

1821. *Bromus lepidus* Holmberg Green

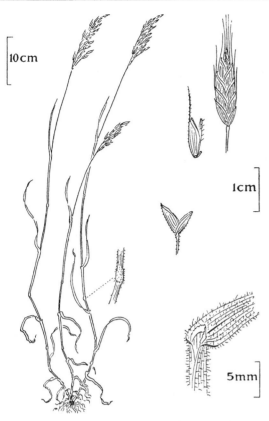

1822.　*Bromus racemosus* L.　Smooth Brome　Green

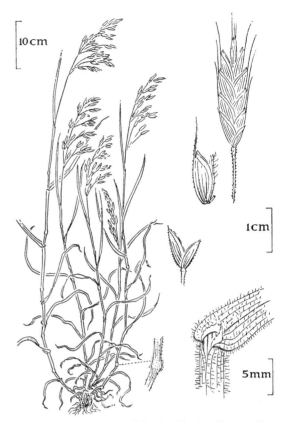

1823.　*Bromus commutatus* Schrad.　Meadow Brome　Green

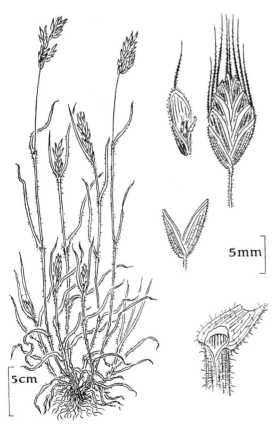

1824.　*Bromus interruptus* (Hack.) Druce　Green

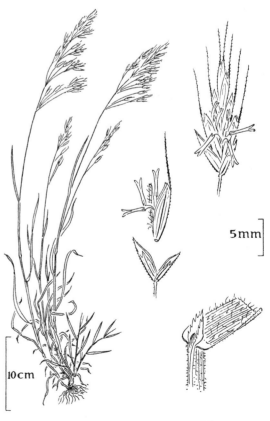

1825.　*Bromus arvensis* L.　Purplish

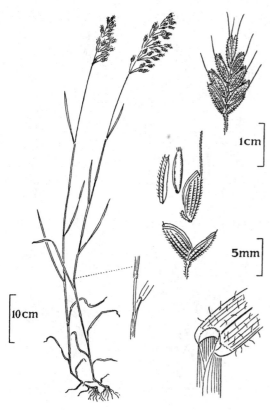

1826. *Bromus secalinus* L. Rye-Brome Green

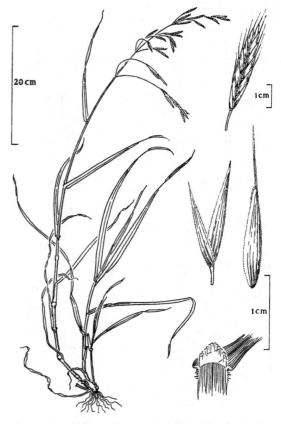

1827. *Ceratochloa carinata* (Hook. & Arn.) Tutin Green

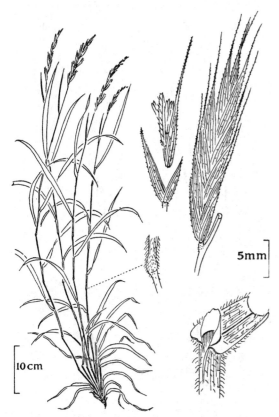

1828. *Brachypodium sylvaticum* (Huds.) Beauv. Slender
False-Brome Green

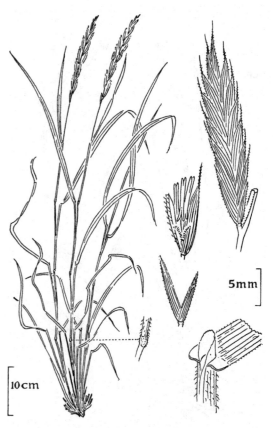

1829. *Brachypodium pinnatum* (L.) Beauv. Heath False-
brome Green

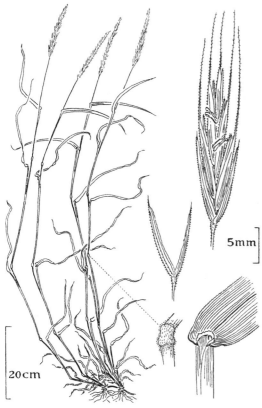

1830. *Agropyron caninum* (L.) Beauv.   Bearded Couch-
grass   Green

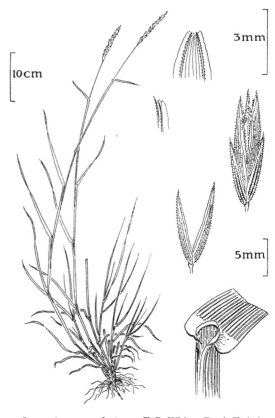

1831. *Agropyron donianum* F. B. White   Don's Twitch
Green or purplish

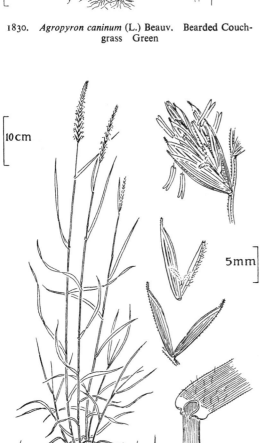

1832. *Agropyron repens* (L.) Beauv.   Couch-grass, Scutch,
Twitch   Green or purplish

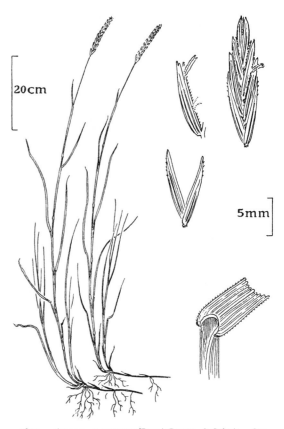

1833. *Agropyron pungens* (Pers.) Roem. & Schult.   Sea
Couch-grass   Greenish

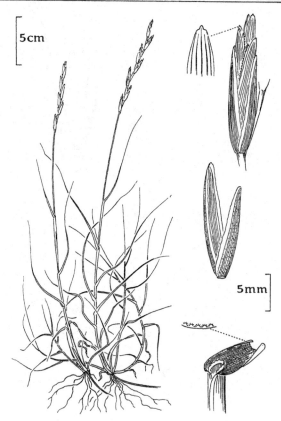

1834. *Agropyron junceiforme* (Á. & D. Löve) Á. & D. Löve
Sand Couch-grass Glaucous

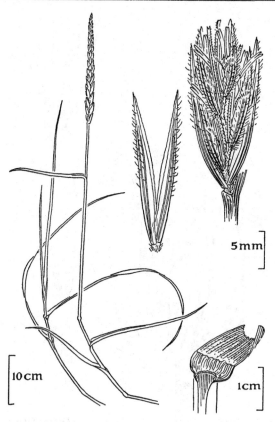

1835. *Elymus arenarius* L. Lyme-grass Glaucous

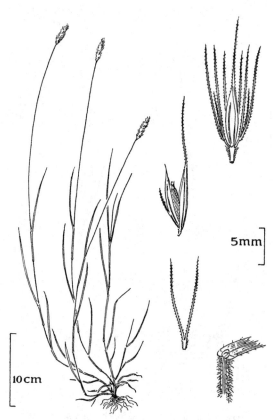

1836. *Hordeum secalinum* Schreb. Meadow Barley Green

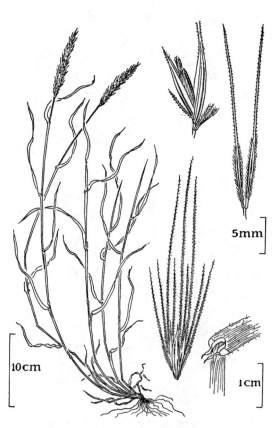

1837. *Hordeum murinum* L. Wall Barley Green

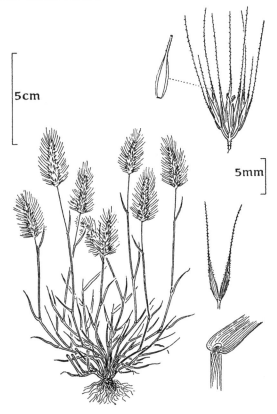

1838. *Hordeum marinum* Huds. Squirrel-tail Grass Green

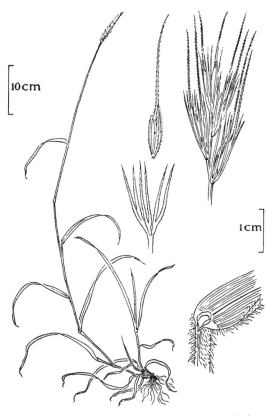

1839. *Hordelymus europaeus* (L.) Harz Wood Barley
Green

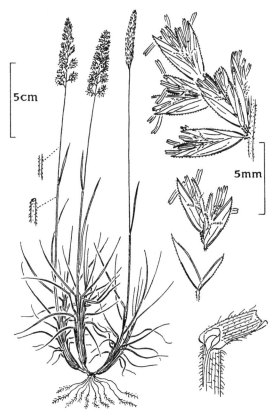

1840. *Koeleria cristata* (L.) Pers. Crested Hair-grass
Purplish, green or whitish

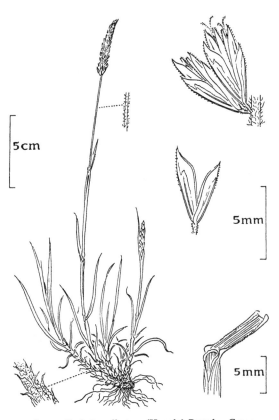

1841. *Koeleria vallesiana* (Honck.) Bertol. Green

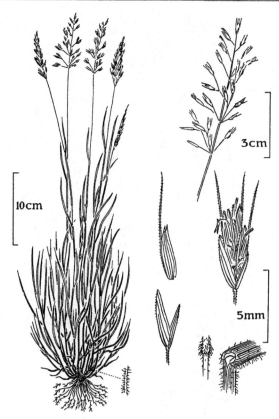

1842.   *Trisetum flavescens* (L.) Beauv.   Yellow Oat
Yellowish

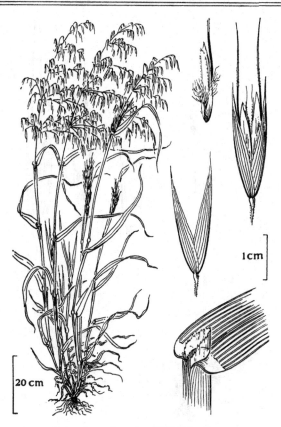

1843.   *Avena fatua* L.   Wild Oat   Green

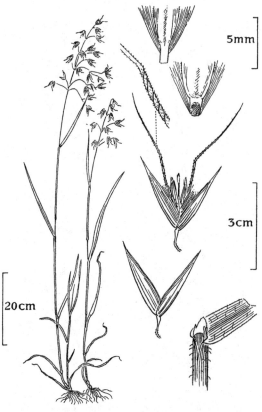

1844.   *Avena ludoviciana* Durieu   Wild Oat   Green

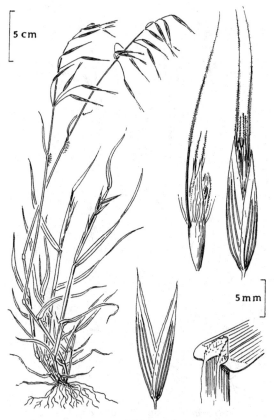

1845.   *Avena strigosa* Schreb.   Black Oat   Green

7·2

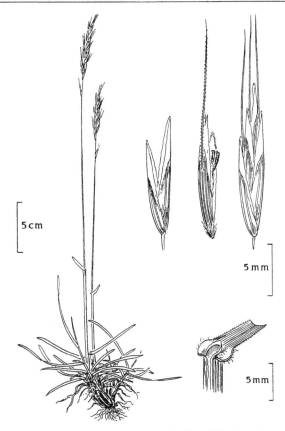

1846.   *Helictotrichon pratense* (L.) Pilger   Meadow Oat
Green or purplish

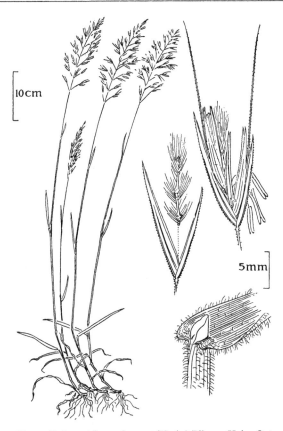

1847.   *Helictotrichon pubescens* (Huds.) Pilger   Hairy Oat
Green or purplish

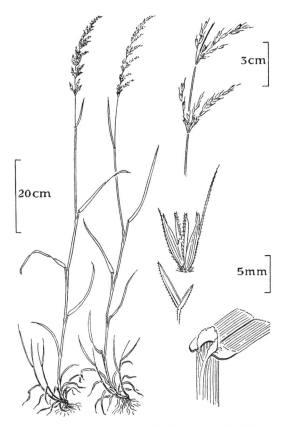

1848.   *Arrhenatherum elatius* (L.) Beauv. ex J. & C. Presl
Oat-grass   Green

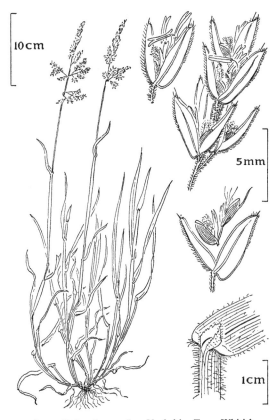

1849.   *Holcus lanatus* L.   Yorkshire Fog   Whitish

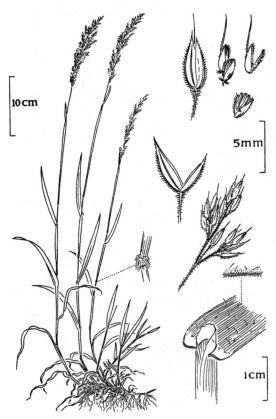

1850. *Holcus mollis* L.   Creeping Soft-grass   Whitish

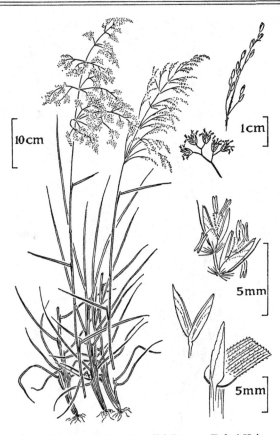

1851. *Deschampsia caespitosa* (L.) Beauv.   Tufted Hair-
grass   Purplish

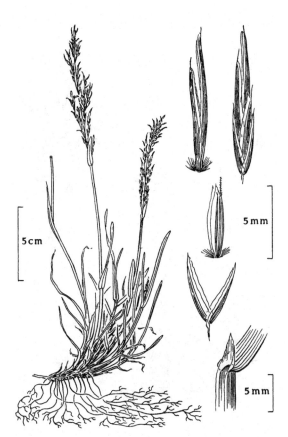

1852. *Deschampsia alpina* (L.) Roem. & Schult.   Purplish

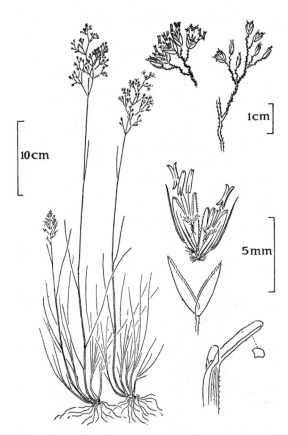

1853. *Deschampsia flexuosa* (L.) Trin.   Wavy Hair-grass
Purplish

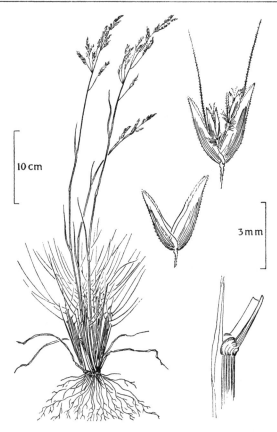

1854.   *Deschampsia setacea* (Huds.) Hack.   Purplish

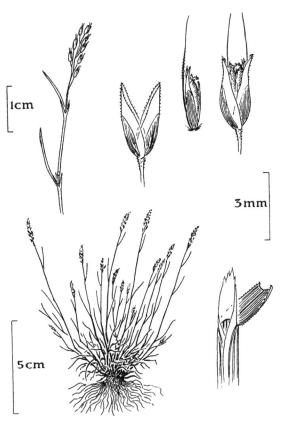

1855.   *Aira praecox* L.   Early Hair-grass   Greenish

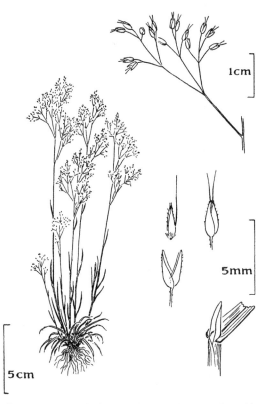

1856.   *Aira caryophyllea* L.   Silvery Hair-grass   Greenish

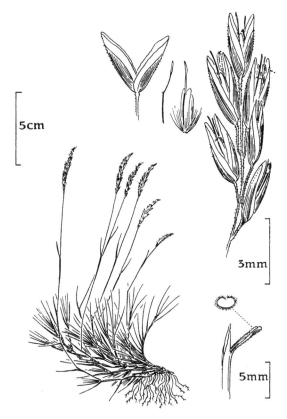

1857.   *Corynephorus canescens* (L.) Beauv.   Purple and
white

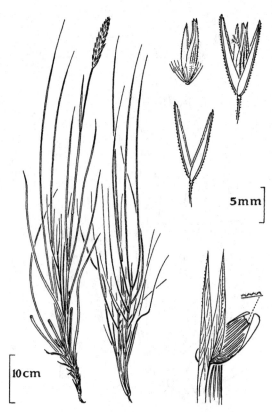

1858.   *Ammophila arenaria* (L.) Link   Marram Grass
Whitish

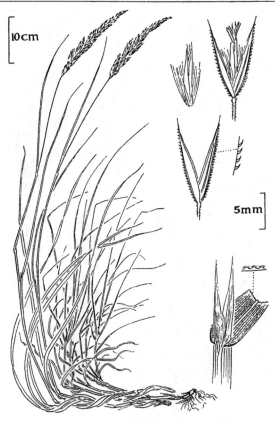

1859.   × *Ammocalamagrostis baltica* (Schrad.) P. Fourn.
Purplish

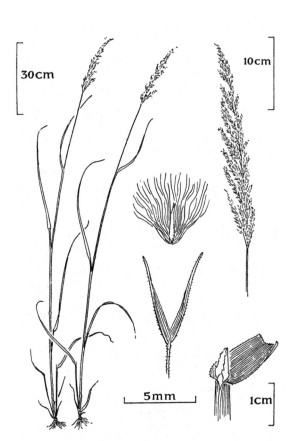

1860.   *Calamagrostis epigejos* (L.) Roth   Bushgrass
Purplish-brown

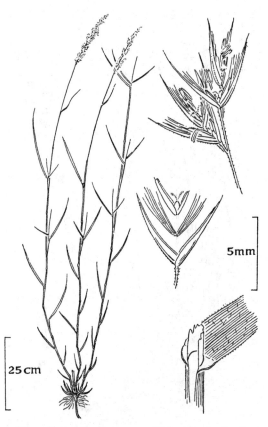

1861.   *Calamagrostis canescens* (Weber) Roth   Purple
Smallreed   Light brown

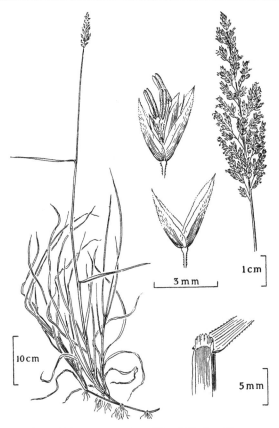

1862. *Calamagrostis stricta* (Timm) Koeler Narrow
Smallreed Brownish

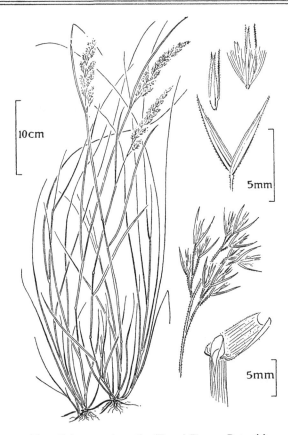

1863. *Calamagrostis scotica* (Druce) Druce Brownish

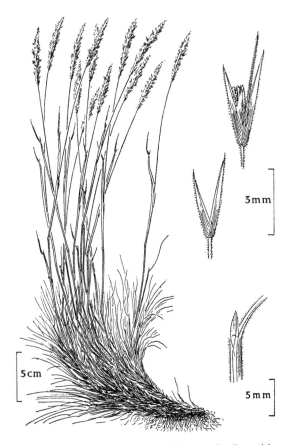

1864. *Agrostis setacea* Curt. Bristle Agrostis Brownish

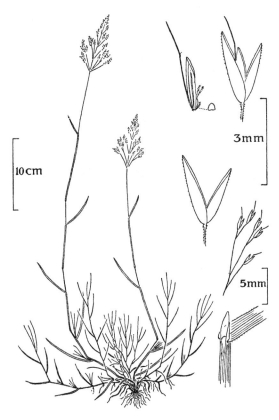

1865. *Agrostis canina* L. Brown Bent-grass Brownish

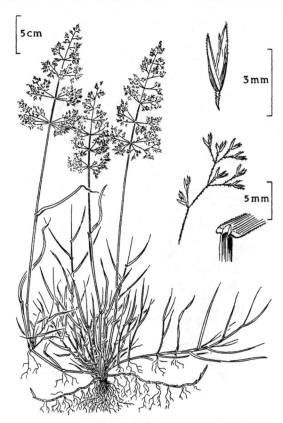

1866. *Agrostis tenuis* Sibth. Common Bent-grass
Brownish

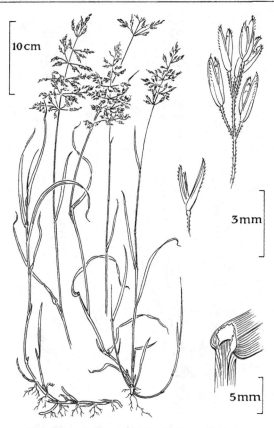

1867. *Agrostis gigantea* Roth   Common Bent-grass
Purplish-brown

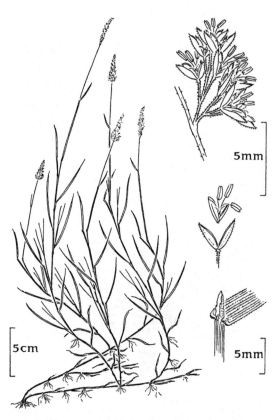

1868. *Agrostis stolonifera* L. Fiorin Brownish

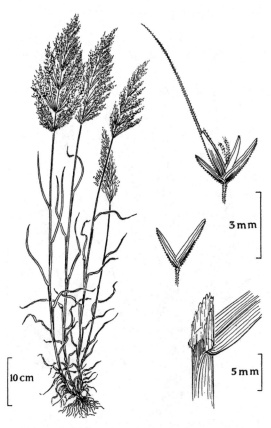

1869. *Apera spica-venti* (L.) Beauv. Silky Apera
Brownish

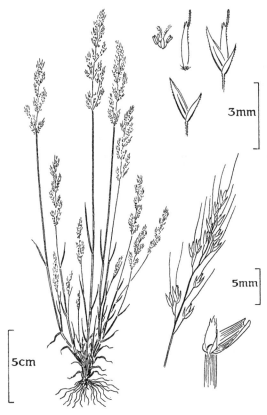

1870.   *Apera interrupta* (L.) Beauv.   Brownish

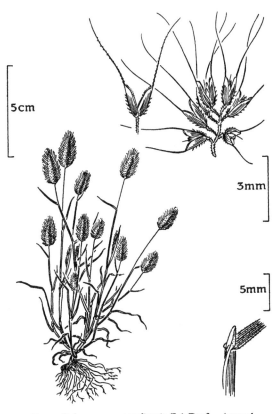

1871.   *Polypogon monspeliensis* (L.) Desf.   Annual
Beardgrass   Greenish

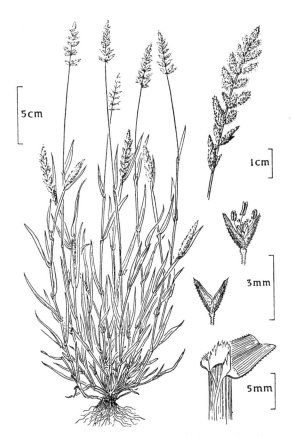

1872.   *Polypogon semiverticillatus* (Forsk.) Hyl.   Beardless
Beardgrass   Greenish

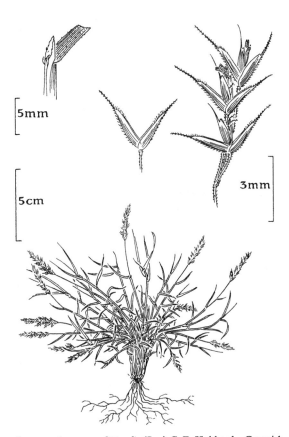

1873.   × *Agropogon littoralis* (Sm.) C. E. Hubbard   Greenish

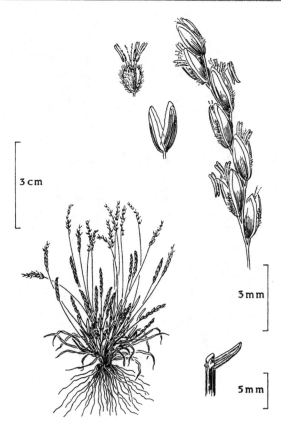

1874. *Mibora minima* (L.) Desv.   Purplish

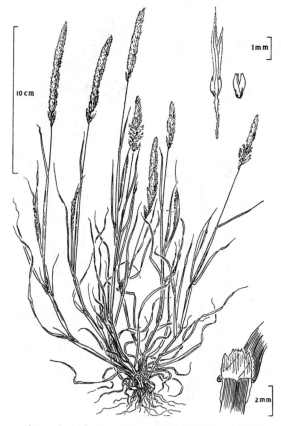

1875. *Gastridium ventricosum* (Gouan) Schinz & Thell.
Nitgrass   Greenish

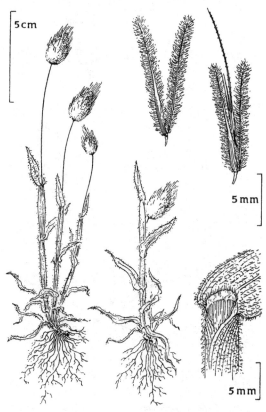

1876. *Lagurus ovatus* L.   Hare's-tail   Whitish

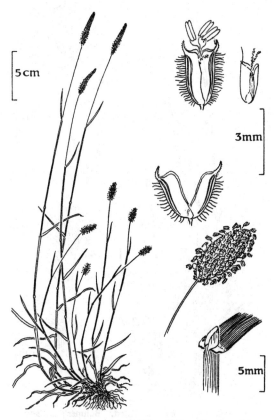

1877. *Phleum bertolonii* DC.   Cat's-tail   Greenish

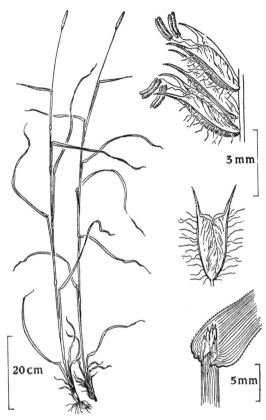

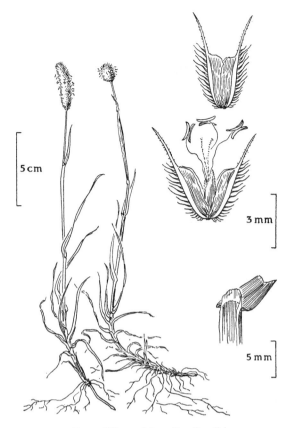

1878.   *Phleum pratense* L.   Timothy   Greenish

1879.   *Phleum alpinum* L.   Purplish

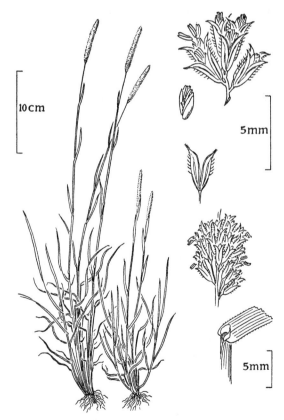

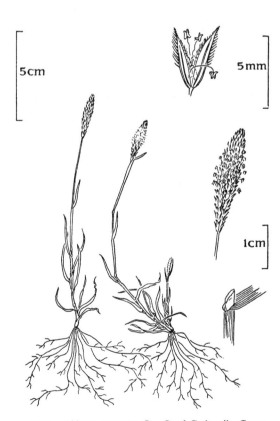

1880.   *Phleum phleoides* (L.) Karst.   Böhmer's Cat's-tail
Green

1881.   *Phleum arenarium* L.   Sand Cat's-tail   Green

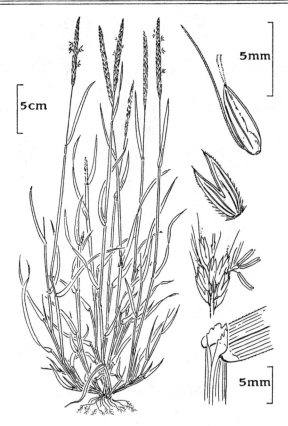

1882. *Alopecurus myosuroides* Huds.   Black Twitch
Greenish or purplish

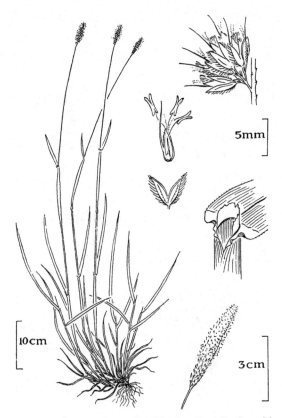

1883. *Alopecurus pratensis* L.   Meadow Foxtail   Greenish

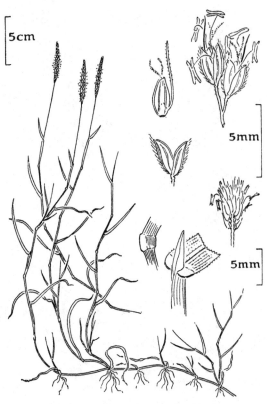

1884. *Alopecurus geniculatus* L.   Marsh Foxtail   Greenish

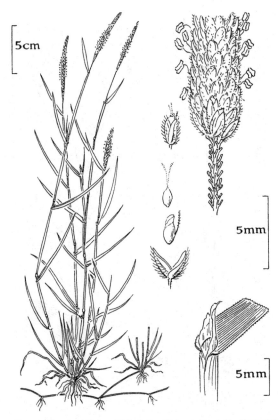

1885. *Alopecurus aequalis* Sobol.   Orange Foxtail   Greenish

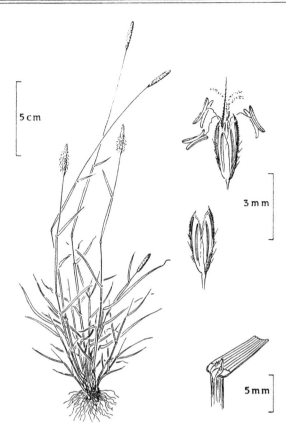

1886.   *Alopecurus bulbosus* Gouan   Tuberous Foxtail
Greenish

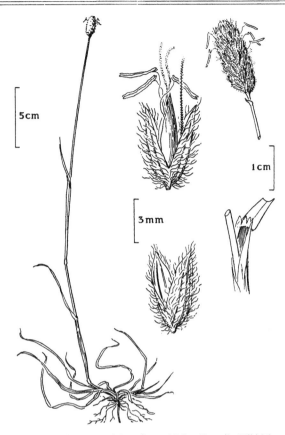

1887.   *Alopecurus alpinus* Sm.   Alpine Foxtail   Whitish

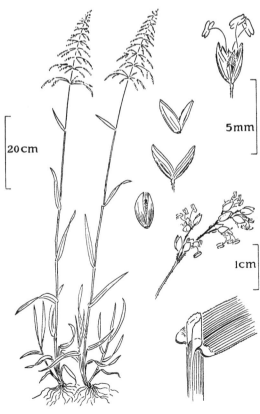

1888.   *Milium effusum* L.   Wood Millet   Green

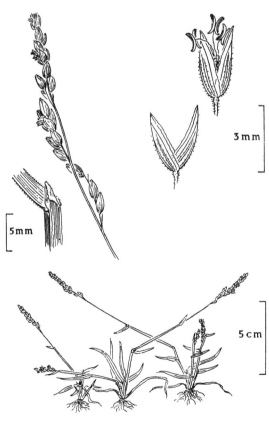

1889.   *Milium scabrum* Rich.   Green

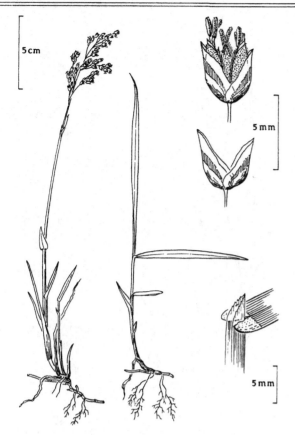

1890. *Hierochloe borealis* (L.) Beauv. Holy-grass
Purplish

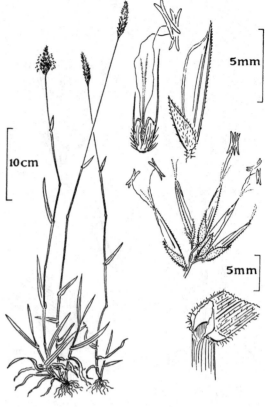

1891. *Anthoxanthum odoratum* L. Sweet Vernal-grass
Greenish

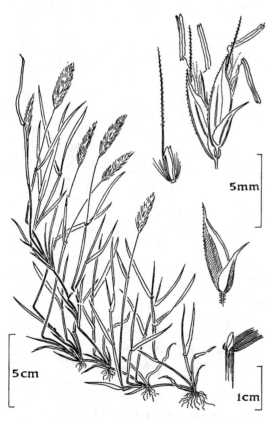

1892. *Anthoxanthum puelii* Lecoq & Lamotte Greenish

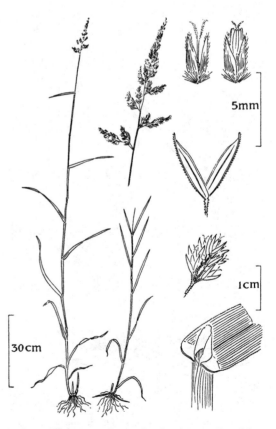

1893. *Phalaris arundinacea* L. Reed-grass Purplish

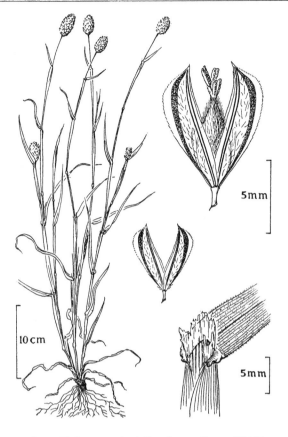

1894. *Phalaris canariensis* L. Canary Grass Whitish

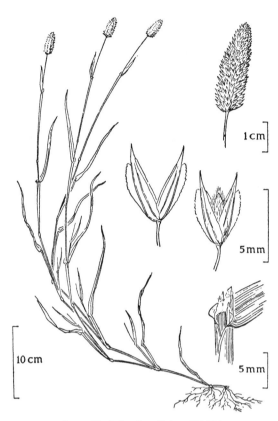

1895. *Phalaris minor* Retz. Whitish

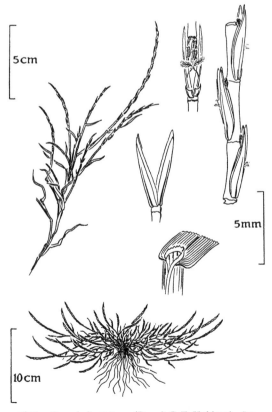

1896. *Parapholis strigosa* (Dum.) C. E. Hubbard Sea
Hard-grass Green

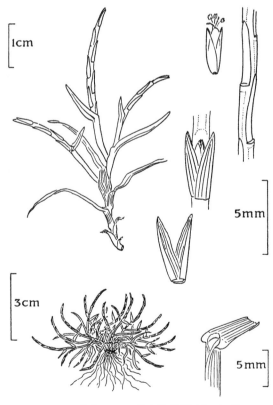

1897. *Parapholis incurva* (L.) C. E. Hubbard Green

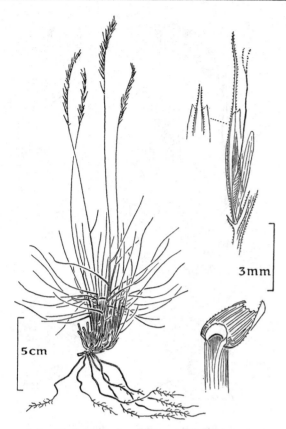

1898. *Nardus stricta* L. Mat-grass Purplish

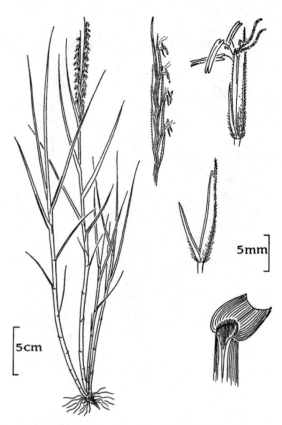

1899. *Spartina maritima* (Curt.) Fernald Cord-grass
Yellowish

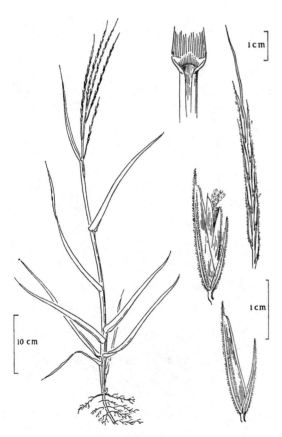

1900. *Spartina × townsendii* H. & J. Groves Cord-grass
Yellowish

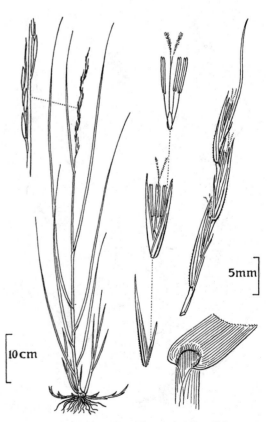

1901. *Spartina alterniflora* Lois. Purplish

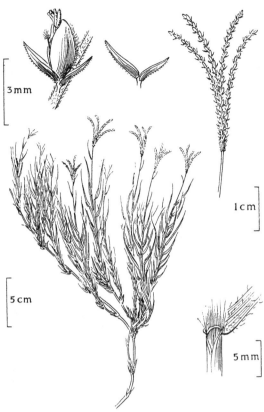

1902.   *Cynodon dactylon* (L.) Pers.   Bermuda-grass
Purplish

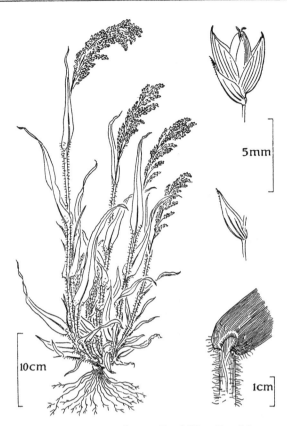

1903.   *Panicum mileaceum* L.   Millet   Purplish

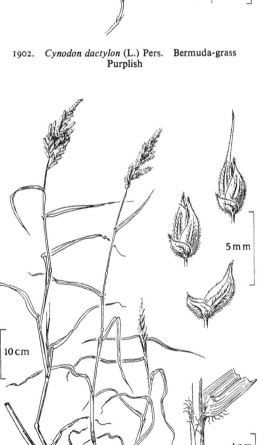

1904.   *Echinochloa crus-galli* (L.) Beauv.   Cockspur
Purplish

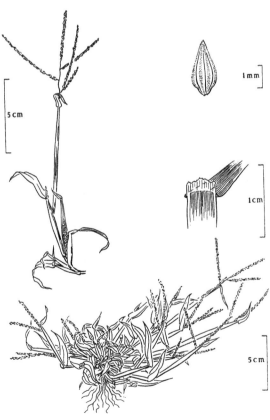

1905.   *Digitaria ischaemum* (Schreb.) Muhl.   Red Millet
Purplish

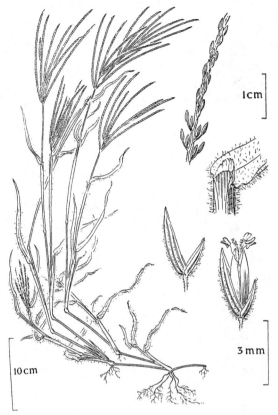

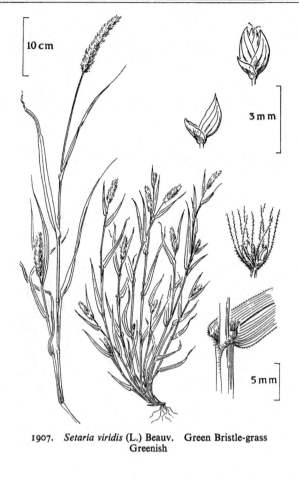

1906. *Digitaria sanguinalis* (L.) Scop.   Crab-grass
Purplish

1907.   *Setaria viridis* (L.) Beauv.   Green Bristle-grass
Greenish

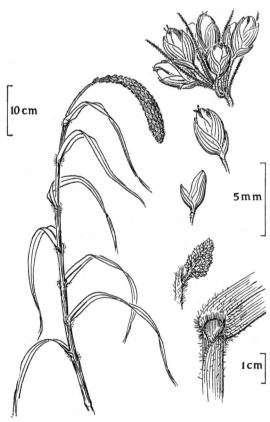

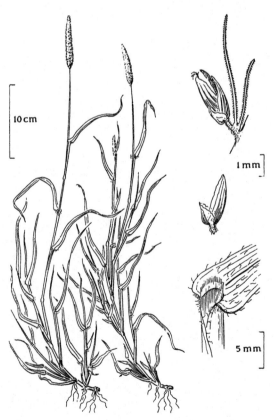

1908.   *Setaria italica* (L.) Beauv.   Millet   Greenish

1909.   *Setaria verticillata* (L.) Beauv.   Greenish

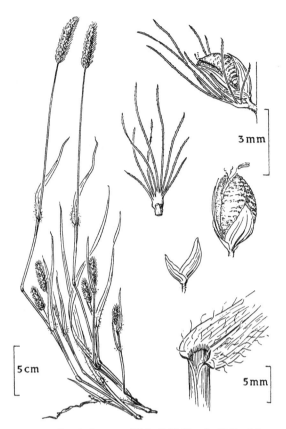

3mm

5cm

5mm

1910.   *Setaria lutescens* (Weigel) Hubbard   Yellowish

INDEX

[117]